T0331770

Plasma Etching Processes for CMOS Device Realization

Series Editor
Robert Baptist

Plasma Etching Processes
for CMOS Device Realization

Edited by

Nicolas Posseme

First published 2017 in Great Britain and the United States by ISTE Press Ltd and Elsevier Ltd

ISTE Press Ltd
27-37 St George's Road
London SW19 4EU
UK

www.iste.co.uk

Elsevier Ltd
The Boulevard, Langford Lane
Kidlington, Oxford, OX5 1GB
UK

www.elsevier.com

Notices

For information on all our publications visit our website at http://store.elsevier.com/

British Library Cataloguing-in-Publication Data
A CIP record for this book is available from the British Library
Library of Congress Cataloging in Publication Data
A catalog record for this book is available from the Library of Congress
ISBN 978-1-78548-096-6

Printed and bound in the UK and US

Contents

Sébastien BARNOLA, Nicolas POSSEME, Stefan LANDIS and Maxime DARNON

Maxime DARNON and Nicolas POSSEME

Preface

Plasma etching is the main technology used to pattern complex structures involving various materials (from metals to semiconductors and oxides to polymers) in domains such as microelectronics, bio-technology, photonics and microsensors. The ability of "cold plasmas", including mainly inductively coupled plasmas (ICP) and capacitively coupled plasmas (CCP), to generate well-controlled etched profiles (anisotropic (vertical) etching) and high etch selectivities (selected material is etched at the much higher rate than others) between materials involved in complex stacks of materials has made this technology successful. The most famous field in which plasma technology has played a key role for the last 40 years is the semiconductor industry.

Plasma etching (using an ionized gas to carve tiny components on silicon wafers) has long enabled the perpetuation of Moore's law (the observation that the number of transistors that can be squeezed into an integrated circuit doubles about every 2 years). Today, etch compensation helps to create devices that are smaller than 20 nm.

Plasma technologies have been critical not only to assist the miniaturization capabilities of lithography with specific processes such as resist trimming processes (lateral erosion of the photoresist mask in order to decrease its critical dimension defined by the lithography) and double patterning (a method of overlaying two patterns to achieve the original design) technologies, but also to pattern complex structures with the appropriate level of dimension control. However, with the constant scaling down in device dimensions and the emergence of planar fully depleted silicon on insulator (FDSOI) or complex 3D structures (like

FinFET sub-20 nm devices, nanowires and stacked nanowires at long term), plasma etching requirements have become more and more stringent with critical dimension control, profile control and film damage at the atomic scale to reach zero variability (precise control of the gate transistor dimension with no damage) required at the horizon of 2020 according to the ITRS.

Now more than ever, plasma etch technology is used to push the limits of semiconductor device fabrication into the nanoelectronics age. This will require improvements in plasma technology (plasma sources, chamber design, etc.), new chemistry methods (etch gases, flows, interactions with substrates, etc.) as well as a compatibility with new patterning techniques such as multiple patterning, EUV lithography, direct self-assembly (DSA), e-beam lithography or nanoimprint lithography.

The goal of this book is to present these etch challenges and associated solutions encountered through the years for transistor realization.

After an introduction to the evolution of CMOS devices through the years in Chapter 1, we will define in Chapter 2 the plasma etching in microelectronics. Then, in Chapter 3, we will present patterning challenges in microelectronics and how plasma etch technology becomes a key solution. Finally, in Chapter 4, we will present the challenges and constraints associated with transistor manufacturing.

Nicolas POSSEME
October 2016

1

CMOS Devices Through the Years

The CMOS transistor is the fundamental building block of modern electronic devices and is ubiquitous in modern electronic systems. Its dimension is typically around 20 nm and the unit cost of the device is around a few nano dollars.

The basic principle of the solid state transistor was stated by Julius Edgar Lilienfeld in 1925 who patented *"a method and apparatus for controlling the flow of an electric current between two terminals of an electrically conducting solid by establishing a third potential between said terminals"* (Figure 1.1) [LIL 25].

The experimental demonstration began in 1947 with American physicists John Bardeen, Walter Brattain and William Shockley who made the first point contact transistor (Figure 1.2). Working at Bell Labs, they were looking for a solution to replace the vacuum tubes which were not very reliable, consumed too much power and produced too much heat to be practical for AT&T's markets. They were jointly awarded the Nobel Prize in Physics in 1956 for their achievement [NOB 47].

Chapter written by Maud VINET and Nicolas POSSEME.

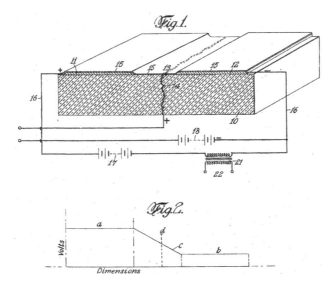

Figure 1.1. *Schematics of Lilienfeld's original patented transistor (from E.J. Lilienfeld [LIL 25])*

Figure 1.2. *Artifact of the first point contact transistor fabricated by Bardeen, Shockley and Brattain in 1947 in Bells Labs*

The first demonstration of a semiconductor-based amplifying device was the "point contact" transistor made out of a germanium crystal lying on a metal plate. The crystal was contacted by a gold strip cut into two pieces which were a hair apart, and covered a triangle which was held in contact by a spring.

To improve the point contact transistor, Shockley conceived the possibility of minority carrier injection and invented an entirely new, considerably more robust, type of transistor with a layered or "sandwich" structure. This structure went on to be used for the vast majority of all transistors in the 1960s, and evolved into the bipolar junction transistor. At that time, though, the junction-based transistors were limited in performance, since they were not able to carry voice signals due to frequency limitations.

The man who paved the way for the improvement was Gordon Teal, who suspected that better grown materials of better quality could lead to better performance. Teal thought transistors should be built from a single crystal, as opposed to cutting a sliver from a larger ingot of many crystals. His method was to take a tiny seed crystal and dip it into the melted germanium then pull slowly as a crystal formed like an icicle below the seed.

In 1959, Dawon Kahng and Martin M. (John) Atalla at Bell Labs invented the metal–oxide–semiconductor field-effect transistor (MOSFET). Operationally and structurally, it was different from the bipolar junction transistor. The MOSFET was made by putting an insulating layer on the surface of the semiconductor and then placing a metallic gate electrode on that. It used crystalline silicon for the semiconductor and a thermally oxidized layer of silicon dioxide for the insulator. The silicon MOSFET did not generate localized electron traps at the interface between the silicon and its native oxide layer, and thus was inherently free from the trapping and scattering of carriers that had impeded the performance of the earlier field-effect transistors.

As the industry expanded, the fundamental fabrication operations (photolithography, etching, deposition and thermal treatment) became more and more specialized and intertwined. Relying on several suppliers, this specialization led the industry to advance at a fast pace. The need for a technology roadmap arose in order to coordinate the industry, so that each supplier could target an appropriate date for their part of the work [GAR 00].

For several years, the Semiconductor Industry Association (SIA) gave the responsibility of coordination to the United States, which led to the creation of an American style roadmap, called the National Technology Roadmap for Semiconductors (NTRS). The SIA produced its first technological roadmap in 1993.

In 1998, the SIA came closer to its European, Japanese, Korean and Taiwanese counterparts by creating the first global roadmap: the International Technology Roadmap for Semiconductors (ITRS). This international group has (as of the 2003 edition) 936 companies which were affiliated to working groups within the ITRS. These companies agreed upon the guidelines for device dimensions and specifications in order to provide guidance to the whole industry [ITR 13].

1.1. Scaling law by Dennard

Moore and Dennard's ideas set the semiconductor industry on a course of developing new integrated circuit process technologies and products on a regular pace and providing consistent improvements in transistor density, performance and power.

Moore's law states that each new generation of process technology was expected to reduce minimum feature size by approximately 0.7x. A 0.7x reduction in the linear features size was considered to be worthwhile for a new process generation as it translated to about a 2x increase in transistor density.

Dennard's scaling law was the supporting physics guidance to preserve the CMOS operation while scaling its dimensions [DEN 74]. The scaling principles described by Dennard and his team were quickly adopted by the semiconductor industry as the roadmap for providing systematic and predictable transistor improvements.

Table 1.1 is reproduced from Dennard's paper and summarizes transistor or circuit parameter changes under ideal scaling conditions, where k is the unitless scaling constant [DEN 74].

Device or Circuit Parameter	Scaling Factor
Device dimension tox, L, W	$1/k$
Doping concentration Na	k
Voltage V	$1/k$
Current I	$1/k$
Capacitance eA/t	$1/k$
Delay time per circuit VC/I	$1/k$
Power dissipation per circuit VI	$1/k^2$
Power density VI/A	1

Table 1.1. *Scaling results for circuit performance (from B. Dennard et al. [DEN 74])*

During the 1970s and 1980s, the semiconductor industry managed to maintain a fast pace and introduced new technology generations approximately every 3 years. This translated into a transistor density improvement of ~2x every 3 years, but this was also a period when average chip size was increasing, resulting in a transistor count increase of close to 2x every 18 months. Starting in the mid-1990s, the semiconductor industry accelerated the pace of introducing new technology generations once every 2 years. Today, the trend of increasing chip size has slowed down slightly due to cost constraints.

Historically, the transistor power reduction afforded by Dennard scaling allowed manufacturers to drastically raise clock frequencies from one generation to the next, without significantly increasing the overall circuit power consumption.

Since around 2005–2007, scaling faced a challenge that broke down Dennard's scaling law. On reaching 100 nm (or the deep sub-micron channel length), maintaining the electrostatic integrity of the transistor became a major issue.

Some of the major problems to MOSFET scaling in sub-100 nm channel length regime were:

1) short channel effect (SCE);

2) drain induced barrier lowering (DIBL);

3) increased off-state current;

4) increased gate leakage;

5) poly gate depletion effects;

6) source/drain access resistance increase;

7) high-field mobility degradation;

8) variability.

In order to overcome these challenges, significant innovations were made at every new technological node.

1.2. CMOS device improvement through the years

The speed of high performance logic circuits depends on the drive current, which is the source–drain saturation current of the metal–oxide–semiconductor field-effect transistor (MOSFET). The improved performance of integrated circuits can be seen from the effect of the scaling on the drive current. Indeed, the saturation current I_{dsat} of the MOSFET can be written as follows:

$$I_{dsat} = \frac{W \times \mu \times C_{ox} \times (V_G - V_T)^2}{2L}$$ [1.1]

where W and L are the width and length of the MOSFET channel, μ is the channel carrier effective mobility, C_{ox} is the gate oxide capacitance, V_G is the gate voltage and V_T is the threshold voltage.

Reduction of channel length L, increase of channel carrier effective mobility μ and gate oxide capacitance density C_{ox} will result in an increased I_{dsat}.

1.2.1. *Mobility improvement*

With the scaling of transistor dimensions, mobility in bulk devices suffered a severe degradation as higher doping levels were required to preserve the CMOS electrostatics. As a result, strain engineering was widely accepted as a promising technique to overcome this mobility degradation and to restore CMOS performance to the 90 nm node.

One key consideration in using strain engineering in CMOS technologies is that PMOS and NMOS respond differently to different types of strain. Specifically, PMOS performance is best served by applying compressive strain to the channel, whereas NMOS benefit from tensile strain (see Figure 1.3). Many approaches to strain engineering induce strain locally, allowing both n-channel and p-channel strain to be modulated independently.

The first approach to be introduced involved the use of a strain-inducing capping layer. Chemical vapor deposition (CVD) silicon nitride was a common choice for a strained capping layer, in that the magnitude and type of strain (e.g. tensile vs. compressive) may be adjusted by modulating the deposition conditions, especially temperature. Standard lithography patterning techniques were used to selectively deposit either tensile or compressive strain-inducing capping layers on either N or PMOS.

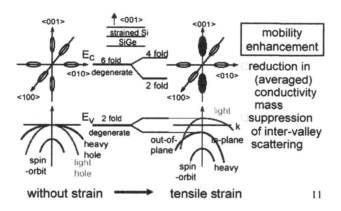

Figure 1.3. *NMOS band diagram with and without tensile strain. Tensile strain induces a degeneracy splitting in the <001> valleys. As a result, the electron concentration increases in the low effective mass subbands, resulting in a lowered averaged conductivity mass (from S-I. Takagi [TAK 03])*

A second prominent approach in strain engineering involves the use of silicon-rich alloys, especially silicon-germanium, to modulate channel strain. For instance, compressive strain can be induced by replacing the source and drain region of a MOSFET with silicon-germanium (Figure 1.4).

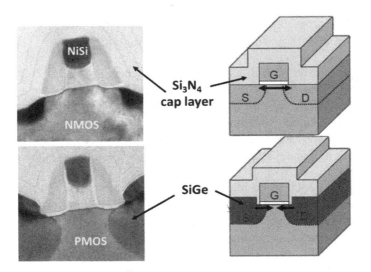

Figure 1.4. *Intel 65 nm technology corresponding to the second generation of uniaxial strained Si for enhanced performance (Lg = 35 nm, Tox = 1.2 nm) (adapted from S. Thompson [THO 05])*

1.2.2. *Leakage current reduction*

The scaling of silicon dioxide dielectrics was an effective approach to enhance transistor performance in complementary metal–oxide–semiconductor (CMOS) technologies as predicted by Moore's law (see equation [1.1]).

In the past few decades, a reduction in the thickness of silicon dioxide gate dielectrics has enabled increased numbers of transistors per chip with enhanced circuit functionality and performance at low costs. However, as the devices approach the sub-45 nm scale, the equivalent oxide thickness (EOT) of the traditional silicon dioxide dielectrics is required to be smaller than 1 nm, which is approximately 3 monolayers and close to the physical limit, resulting in high gate leakage currents due to the obvious quantum tunneling effect at this scale.

Figure 1.5 shows how gate leakage becomes the main contributor to leakage current when the equivalent oxide thickness was scaled down in the 1.5 nm range.

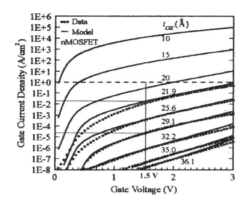

Figure 1.5. *Gate leakage as a function of gate voltage for SiO₂ gate oxide thicknesses ranging from 3.6 nm down to 1 nm. The threshold of 1 A/cm² was reached when gate oxide was below 2 nm for V_g = 1.5 V (from Y. Taur and E.J. Nowak [TAU 97])*

To continue the downscaling of the EOT, dielectrics with a higher dielectric constant (high-k) were suggested as a solution to achieve the same transistor performance and electrostatics, while maintaining a relative physical thickness.

Many candidates with a probable high-k gate dielectrics were considered to replace SiO_2 (see Table 1.2).

Dielectric	Dielectric constant
SiO_2	3.9
Si_3N_4	7
Al_2O_3	9
Y_2O_3	15
La_2O_3	30
Ta_2O_5	26
TiO_2	80
HfO_2	25
ZrO_2	25

Table 1.2. *High-k gate dielectric candidates to replace SiO₂*

The industry first introduced oxynitride gate dielectrics in the 1990s, wherein a conventionally formed silicon oxide dielectric was infused with a small amount of nitrogen. The nitride content subtly raised the dielectric constant and allowed for a relaxed gate oxide thickness while maintaining good electrostatics.

To further contain the gate leakage, the gate oxide was afterwards replaced with high-k (high dielectric constant) dielectric material based on hafnium oxides (HfSiON and HfO_2) and the gate leakage was divided by 4 decades, as illustrated in Figure 1.6. Intel was the first company to announce the deployment of high-k dielectrics in its 45 nm bulk technology.

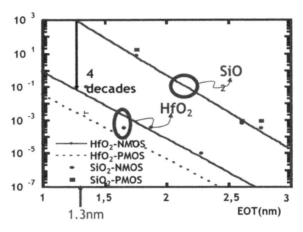

Figure 1.6. *Gate leakage as a function of gate voltage for SiO_2 and HfO_2 gate oxides. The effectiveness of HfO_2 introduction is illustrated by a 4-decades reduction of the gate current at a given equivalent oxide thickness (EOT) as compared to SiO_2 reference (from B. Guillaumot et al. [GUI 02])*

Together with the introduction of high-k dielectrics, metal gate stacks based on titanium nitride (TiN) replaced the doped polysilicon gate. A major challenge in the introduction of metal gate electrodes was the need to obtain distinct gate work functions for NMOS and PMOS devices. While two metals would ordinarily be required, a method was developed that allowed the metal gate work function to be tuned over the required range, based on Al and La diffusion in the gate dielectric.

The introduction of metal gate stack suppressed the polydepletion effect, as illustrated in Figure 1.7. Polydepletion was on the way to benefiting from thin equivalent oxide thickness.

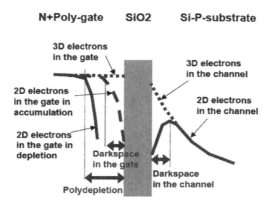

Figure 1.7. *Schematics of the polydepletion effect in a N+ (for the sake of the example) polysilicium gate. This polydepletion results in an apparent increase of the gate oxide thickness, resulting in a lower electrostatic control of the gate over the channel*

1.2.3. *Gate-last approach*

Concomitantly with the high-k/metal gate stack introduction, the gate stack integration scheme was changed from gate-first to gate-last to ensure low Vt PFETs. Indeed, the P-type work-function tends to get more and more mid-gap as the metal undergoes thermal budgeting. Thus, in order to target low PFET Vt, the metal gate stack is deposited after junction activation. The gate-last process, also known as the replacement metal gate process (RMG), consists of a Damascene process where a dummy gate is used to define a cavity which will be emptied after RSD definition (see [GUI 02] for early work).

Figure 1.8 represents a simplified scheme of the gate-last approach. After dummy gate patterning using the conventional approach and spacer realization (Figure 1.8(a)), a silicon dioxide film is deposited and planarized by chemical mechanical polishing (Figure 1.8(b)). In Figure 1.8(c), the dummy gate is selectively removed, creating a cavity in the silicon dioxide. Finally, high-k–work-function metal (TiN) (Figure 1.8(d)) and capping metal gate (tungsten) are deposited (Figure 1.8(e)) followed by a planarization step (Figure 1.8(f)).

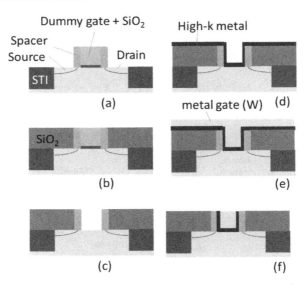

Figure 1.8. *Gate-last integration scheme*

1.2.4. *New transistor generations*

1.2.4.1. *FinFET*

The most dramatic change in CMOS transistors was brought by Intel at 22 nm node, in the year 2011, with the introduction of FinFET architecture in production. FinFET devices belong to the fully depleted architectures (sometimes this statement can be argued depending on the junction isolation to the bulk substrate) like FDSOI, Gate-all-Around transistors and all the architectures where we can define a silicon thickness or width that is smaller than the depletion depth in the device. The architectures of the fully depleted family are shown in Figure 1.9.

The fully depleted architectures were introduced to restore the gate electrostatic control over the channel. Indeed when the gate length became shorter than 30 nm, it becomes extremely challenging for bulk planar devices to maintain low I_{off} currents, in spite of very high channel doping and the use of halos (highly doped pockets at the entrance of the channel, of opposite channel type, designed in order to enhance the potential barrier at the source side).

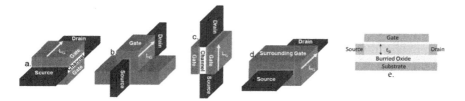

Figure 1.9. *(a) Planar DG [BAL 87], (b) FinFET [HIS 89], (c) vertical surrounding gate [TAK 88], (d) gate all around (GAA) [COL 90], (e) thin film transistor on top of a buried oxide with a ground plane (from M. Fukuma [FUK 88])*

Electrostatics in fully depleted architectures is controlled by the thin silicon film channel thickness or the Fin width. As shown in equation [1.2], the DIBL (and subthreshold slope) improves for a given gate length as a function of thinner silicon channel thickness or thinner Fin width as implemented in the Mastar model [SKO 93], which was the analytical model used to build ITRS specifications:

$$DIBL = \frac{\varepsilon Si}{\varepsilon Ox} \frac{Tinv.TSi}{Lg^2} Vds \qquad [1.2]$$

Figure 1.10 provides some SEM observations of Intel 22 nm FinFET. The fact that the active width is now discrete (a multiple of the number of Fin) appears clearly.

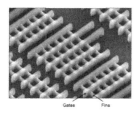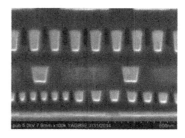

Figure 1.10. *Left: top-down SEM observation of Intel 22 nm FinFET technology (from http://www.intel.com/content/www/us/en/silicon-innovations/intel-22nm-technology.html). Middle: 14 nm Fin cross-section. Right: first metal levels cross-section (from S. Natarajan [NAT 14])*

1.2.4.2. *Fully depleted silicon on insulator (FDSOI)*

Another answer from the semiconductor industry to electrostatics recovery was the introduction of FDSOI technology by STMicroelectronics at 28 nm node. FDSOI is a planar technology and is fabricated using a thin Si film (less than 10 nm) on a buried oxide insulator (BOX). In recent decades, FDSOI has been demonstrated to offer additional features and benefits compared to the other options including: i) total dielectric isolation to lower junction leakage, capacitance and latch-up immunity; ii) undoped channel to reduce threshold voltage variation and enable higher mobility; iii) ultra-thin BOX (around 25 nm) to improve electrostatics, and enable back-biasing for Vt tuning and power/performance trade-off optimization; iv) multiple work-function metal gates and ground-plane doping for multi-Vt devices; v) simple co-integration of bulk and FDOI devices allowing for legacy bulk IP preservation; vi) simple planar layout similar to conventional bulk technology, allowing for re-use of most of the previous generation bulk CMOS FEOL modules and simple migration of digital libraries and designs to FDSOI [NGU 14]. ARM-based cores operating up to a record frequency of 3 GHz [LEP 13, MAG 13] have been demonstrated at the system level using 28 nm node ground rules [PLA 12]. FDSOI technologies have begun to reach mainstream manufacturing. The marketplace presence of FDSOI is further expanding with foundry, partners, ecosystem and IP providers.

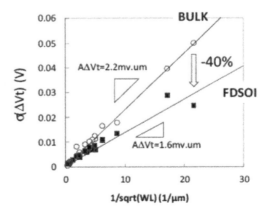

Figure 1.11. *Vt matching as a function of the square root of device area. Because of random dopant fluctuation suppression in the FDSOI device, the variability of small devices is reduced by 40% as compared to the bulk reference (from N. Planes [PLA 12])*

One last point to be highlighted is the fact that both FinFET and FDSOI have undoped or low doped channels. In turn, it translates to an improved variability with respect to bulk technology due to the suppression of random dopant fluctuations (RDF), as shown in Figure 1.11. The variability of small devices is reduced by 40% in FDSOI as compared to bulk devices. This large gain has restored SRAM functionality and allowed for lowered Vmin.

Figure 1.12 provides a SEM observation of STMicroelectronics 28 nm FDSOI [BOE 13].

Figure 1.12. *SEM observation of STMicroelectronics 28 nm FDSOI technology (from Boeuf [BOE 13])*

1.2.5. Patterning challenges for FinFET and FDSOI

Although the FinFET fabrication process reuses a large part of the well-established conventional CMOS processes [JUR 09], the patterning of the fin (FinFET active area) and the gate requires a much tighter process control than for their planar counterpart. While for the FDSOI device, the integrity (limitation of silicon consumption without damage) of the SOI is key after gate patterning.

Among these processes, plasma etch technology is key to the success of device realization today. Indeed, plasma etching, used for more than 40 years in semiconductors manufacturing, confronts the fundamental limits of physics and chemistry. Plasma etching has long enabled the perpetuation of Moore's law. Without the compensating capabilities of plasma etching, Moore's law would have faltered around 1980 when transistor sizes were about 1 micron.

For a sub-20 nm gate length, a fin width of about 10 nm is required, with a width uniformity of about 1 nm, to provide accurate threshold voltage V_T control. The patterning of the fin which was way below the lithographic resolution was for the first time the smallest feature at the tightest pitch. In this case, plasma technologies have been critical to assist the miniaturization capabilities of the lithography with specific processes such as resist trimming processes (lateral erosion of the photoresist mask in order to decrease its critical dimension defined by the lithography) [RAM 02] or double patterning (a method of overlaying two patterns to achieve the original design) [ZIM 09]. Today, etch compensation helps to create devices that are smaller than 20 nm.

Today, plasma technologies are critical not only to assist the miniaturization capabilities of lithography, but also to pattern complex structures with the appropriate level of dimension control. With the constant scaling down in device dimensions and the emergence of planar fully depleted silicon on insulator (FDSOI) or FinFET sub-20 nm devices, plasma etching requirements become more and more stringent with critical dimension control, profile control, film damage at the atomic scale (as illustrated Figure 1.13) to reach zero variability (precise control of the fin and gate transistor dimension with no damage) required at the horizon of 2020 according to the ITRS.

Now more than ever, plasma etch technologies are used to push the limits of future semiconductor device fabrication. New chemistries and new etch tools will have to be developed and proposed to overcome the current patterning issues.

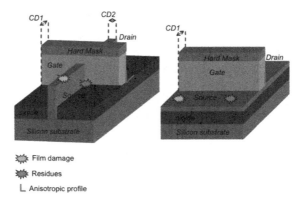

Figure 1.13. *Plasma etch challenges for FinFET or FDSOI device realization*

1.3. Summary

To finish the walk through modern-day devices, Figure 1.14 summarizes the typical dimensions of the technologies currently in production or development. Figures 1.14(b) and (c) provide assumptions on the technology definition for generations to come. It is interesting to keep in mind that 80 nm is the minimum pitch reachable by single exposition in an immersion deep UV stepper, that the 40 nm wire pitch forecasted for 7 nm node corresponds to the limit of double exposure. With the current Si CMOS devices, today it sounds difficult to go below a 40 nm gate pitch because of the electrostatics of the devices: in 40 nm, we have to fit a contact, a gate and two spacers. 24 nm FinFET pitch sounds as a limit because of the Fin patterning challenges, and finally M1 pitch scaling does suffer from the RC limits of Cu wires and does not currently make sense to be decreased below 24 nm.

(a)

Company		28nm	22nm	20nm	16nm
Intel	LG		24		
	Fin pitch		60		
	CPP		90		
	M1		80		
	Year Prod		2011		
Samsung	LG				
	Fin pitch	NA		NA	
	CPP		113		90
	M1		90		64
	Year Prod		2011		2013
TSMC	LG				33
	Fin pitch	NA		NA	45
	CPP		118	90	90/(80)
	M1		90	64	64
	Year Prod		2011	2013	2015
GF	LG				
	Fin pitch	NA	NA		
	CPP				
	M1				
	Year Prod			2016	
IBM	LG				
	Fin pitch				
	CPP				
	M1				
	Year Prod				
ST	LG		24		
	Fin pitch	NA			
	CPP		114		
	M1		90		
	Year Prod		2015		

(b)

Node	CGP	Lgate	Transistor
40-45	170-180	40-45	Planar
28-32	120-130	30-35	Planar
22-14	70-100	25-30	FinFET
10	60-65	20-23	FinFET
7	45-55	12-18	FinFET
5	30-40	10-15	??
3	25-35	7-12	??

(c)

Foundry Node Name	N14	N10	N7	N5	N3
Wire Pitch (nm)	64	48	40	32	24
Contacted Poly Pitch (nm)	80	60	50	40	30
Fin pitch (nm)	48	36	30	24	48

Figure 1.14. *(a) Typical dimensions of technologies from 28 down to 16 nm extracted from supplier websites. Year Prod = Year of introduction for volume production. LG = Min Gate Length. Fin Pitch: when marked NA means planar technology (bulk or FDSOI). All dimension in nm. (b) Projected dimensions for gate length and contacted gate pitch (CGP) as seen from IBM adapted from [HOO 16]. (c) Tentative technology dimensions definition based on Design Technology Co-Optimization(DTCO) rules (from L. Liebmann et al. [LIE 16])*

1.4. What is coming next?

Today, in 2016, all the dimensions we are dealing with at the front end level are below a 100 nm in the most advanced CMOS technology. Typical Fin width is in the 10 nm range, channel thickness of a FDSOI transistor is around 5 nm and gate length is below 30 nm. As compared to the early age of Si CMOS, we count more than 20 chemical elements (see Figure 1.15) in devices nowadays.

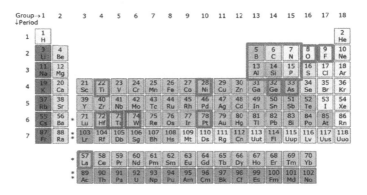

Figure 1.15. *Mendeleev's table highlighting the chemical elements present in FEOL and MEOL of nowadays devices*

Nobody knows for sure what is going to come next. Stacked silicon nanowires are a serious contender to pursue scaling, making it all the more important to preserve the original channel material all along the process. The semiconductor industry observes a lot of excitement around 2D transition metal dichalcogenides TMDs, MX_2 where M is a metal (Mo, W, Hf, etc.) and X is a chalcogenide (like S or Se). The TMDs exhibit strong modulation of their electro-optical properties depending on their composition (gap ranging from 0.4 to 1.8 eV being direct or indirect) and possess the ability to be stacked on each other. They are seen as ideal candidates for aggressive miniaturization of field-effect transistors (FETs) to the single digit nanometer scale because of their ultra-small body thickness which leads to very small electrostatic characteristic scaling length, $\lambda = \sqrt{(\varepsilon_s.t_s.t_{ox}/\varepsilon_{ox})}$.

Beyond the limitations of physics solved by the introduction of new materials and new architectures, the realization of advanced devices is facing the limitation of patterning. Among the pattering steps, plasma etch technology has enabled continuous downscaling in the semiconductor industry for more than 40 years. But this technology is now facing its own issues. Before describing these issues and associated solutions (Chapters 3 and 4), let us define first: plasma etching for microelectronics.

1.5. Bibliography

[BAL 87] BALESTRA F., CRISTOLOVEANU S., BENACHIR M. *et al.*, "Double-gate silicon-on-insulator transistor with volume inversion: a new device with greatly enhanced performance", *IEEE Electron Device Letters*, vol. 8, no. 9, pp. 410–412, 1987.

[BOE 13] BOEUF F., "Ultra Thin Body and Box-FDSOI (UTBBFDSOI): from Research to Industrialization (2000–2012)", *LETI Innovation Days*, available at: www.leti-innovationdays.com, July 2013.

[COL 90] COLINGE J.P., GAO M.H., ROMANO-RODRIGUEZ A. *et al.*, "Silicon-on-insulator 'gate-all-around device'", *Electron Devices Meeting, IEDM '90. Technical Digest., International*, pp. 595–596, 1990.

[DEN 74] DENNARD R.H., GAENSSLEN F., HWA-NIEN Y. *et al.*, "Design of ion-implanted MOSFET's with very small physical dimensions", *IEEE Journal of Solid State Circuits*, vol. SC–9, no. 5, 1974.

[FUK 88] FUKUMA M., "Limitations on MOS ULSIs", *Symposium on VLSI Technology*, p. 7, 1988.

[GAR 00] GARGINI P., "The International Technology Roadmap, for Semiconductors (ITRS): past, present and future", *22nd Annual Gallium Arsenide Integrated Circuit (GaAs IC) Symposium, IEEE*, pp. 3–5, 2000.

[GUI 02] GUILLAUMOT B., GARROS X., LIME F. *et al.*, "75 nm damascene metal gate and high-k integration for advanced CMOS devices", *IEDM*, pp. 355–358, 2002.

[HIS 89] HISAMOTO D., KAGA T., KAWAMOTO Y. *et al.*, "A fully depleted lean-channel transistor (DELTA)-a novel vertical ultra thin SOI MOSFET", *Electron Devices Meeting, Technical Digest., International*, pp. 36–39, 1989.

[HOO 16] HOOK B.T., DORIS B., GOPALAKRISHNAN K., "The Future of HP Computing – Technology Scaling and Hardware Accelerators", *Short Course, Symposium on VLSI Technology*, 2016.

[ITR 13] ITRS, ITRS 2013 Edition, available at: http://itrs.net, 2013.

[JUR 09] JURCZAK M., COLLAERT N., VELOSO A. et al., "Review of FinFET technology", *IEEE International SOI Conference*, 2009.

[LEP 13] LE PAILLEUR L., "Fully-depleted-silicon-on-insulator: from R&D concept to industrial reality", *SOI-3D-Subthreshold Microelectronics Technology Unified Conference* (S3S), *IEEE*, 2013.

[LIE 16] LIEBMANN L., ZENG J., ZHU X. et al., "Overcoming scaling barriers through design technology co-optimization", *Symposium on VLSI Technology*, 2016.

[LIL 25] LILIENFELD E.J., A method and apparatus for controlling electric currents, US 1745175, 22 October 1925.

[MAG 13] MAGARSHACK P., FLATRESSE P., CESANA G., "UTBB FD-SOI: a process/design symbiosis for breakthrough energy-efficiency", *Design, Automation & Test in Europe Conference & Exhibition*, 2013.

[NAT 14] NATARAJAN S., AKBAR S., BOST M. et al., "A 14 nm logic technology featuring 2nd-generation FinFET, air-gapped interconnects, self-aligned double patterning and a 0.0588 μm^2 SRAM cell size", *IEEE International Electron Devices Meeting*, pp. 71–73, 2014.

[NOB 47] NOBEL PRIZE, "The Nobel Prize in Physics 1956", available at: nobelprize.org, accessed 7 December 2014.

[NGU 14] NGUYEN B.-Y., ALLIBERT F. et al., "Fully depleted SOI technology overview", in BROZEK T. (eds), *Micro- and Nanoelectronics Emerging Device Challenges and Solutions*, CRC Press, 2014.

[PLA 12] PLANES N., WEBER O., BARRAL V. et al., "28 nm FDSOI technology platform for high-speed low-voltage digital applications", *Symposium on VLSI Technology*, pp. 133–134, 2012.

[RAM 02] RAMALINGAM S., LEE C., VAHEDI V., "Photoresist trimming: etch solutions to CD uniformity and tuning", *Semiconductor International*, September 2002.

[SKO 93] SKOTNICKI T., MERCKEL G., DENAT C., "MASTAR – A model for analog simulation of subthreshold, saturation and weak avalanche regions In MOSFETs", *International Workshop on VLSI Process and Device Modeling*, pp. 146–147, 1993.

[TAK 88] TAKATO H., SUNOUCHI K., OKABE N. et al., "High performance CMOS surrounding gate transistor (SGT) for ultra high density LSIs", *Electron Devices Meeting*, pp. 222–225, 1988.

[TAK 03] TAKAGI S., "Strained Silicon Technology", Short course IEDM 2003.

[TAU 97] TAUR Y., NOWAK E.J., "Self-aligned (top and bottom) double-gate MOSFET with a 25 nm thick silicon channel", *IEDM*, pp. 427–430, 1997.

[THO 05] THOMPSON S., CHAU R.S., GHANI T. *et al.*, "In search of "Forever", continued transistor scaling one new material at a time", *IEEE Transactions on Semiconductor Manufacturing*, vol. 18, no. 1, 2005.

[ZIM 09] ZIMMERMAN P., "Double patterning lithography: double the trouble ordouble the fun?", *SPIE Newsroom*, 2009.

2

Plasma Etching in Microelectronics

Plasma etching technologies have been used for many years in the semiconductor industry to transfer the patterns defined by lithography in the active materials that form the transistors and the interconnects. More recently, plasma etching processes have also participated in the race towards smaller dimensions, first by helping to shrink the dimensions of the patterns to dimensions smaller than the lithography resolution (so-called trimming or shrinking processes), then by helping to reduce the sidewall roughness of the patterns, and finally by enabling multiple patterning.

In this chapter, we will briefly explain what are the plasma used in plasma etching and describe the standard etch tools used in the semiconductor industry. Then, we will discuss the plasma/surface interactions that lead to plasma etching and the fundamental mechanisms involved in the transfer of patterns during plasma etching. Finally, we will describe the typical pattern profiles that can be observed after plasma etching.

2.1. Overview of plasmas and plasma etch tools

2.1.1. *Overview of plasmas*

According to the Oxford English dictionary, a plasma is defined as "*An ionized gas containing free electrons and positive ions, formed usually at very high temperatures (as in stars and in nuclear fusion experiments) or at*

Chapter written by Maxime DARNON.

low pressures (as in the upper atmosphere and in fluorescent lamps); esp. such a gas which is electrically neutral and exhibits certain phenomena due to the collective interaction of charges". Plasmas can be classified into two main families: fully ionized plasmas where the ions and electrons are close to thermal equilibrium, which are referred to as hot plasmas, and weakly ionized plasmas operating in non-equilibrium, which are referred to as cold plasmas. Only cold plasmas that are used in the semiconductor industry for plasma etching will be considered here.

A plasma can be characterized by many parameters. The plasma electron density n_e corresponds to the density of electrons in the plasma. It varies between $\sim 10^8$ and $\sim 10^{12}$ cm^{-3} for etch applications. The plasma ionization degree is defined as the proportion of positive ions in the plasma: $\alpha = \dfrac{n_+}{n_+ + n_n}$, with n_+ the density of positive ions and n_n the density of neutral species. For typical etch plasmas, $\alpha \ll 1$. The electron temperature T_e corresponding to the average electron energy, is close to 3 eV for etch plasmas. The plasma potential is close to 5 T_e and is therefore close to 15 eV. The ion temperature is close to the neutral species temperature, which ranges from room temperature to ~600 K (0.05 eV). The operating pressure for etched plasmas ranges from few millitorrs to few hundreds of millitorrs. At room temperature, a millitorr corresponds to a density of $\sim 3.10^{13}$ cm^{-3}. During plasma processes, most of these plasma parameters are not known.

One of the specificities of plasmas is their quasi-neutrality. This means that the bulk of the plasma contains a similar amount of positive and negative charges. In other words, the average flux of positive charges and negative charges leaving the plasma must be identical. On the other hand, electrons are much more energetic and mobile than ions, and would therefore escape the plasma faster than the ions. To equate the flux of ions and electrons, the plasma charges itself to a positive potential (so-called the plasma potential V_P) compared to its boundaries. As a result, a space charge region called the sheath builds up between the plasma and all the surfaces. Positive charges are accelerated in the sheath outwards from the plasma while the negative charges (especially the electrons) are repelled into the plasma.

Therefore, ions bombard all the surfaces and reach the surface with a preferential direction (perpendicular to the surface) that can be used to make the plasma processes anisotropic.

Even if plasmas operate in a non-equilibrium state, the various species in the plasma can be considered to be at equilibrium with itself. As a result, the electron energy distribution function (EEDF) is a Maxwellian function, and the average electron energy defined as the electron temperature is around 3 eV for etching plasmas. Electrons can collide with the species from the plasma but the large difference in mass between electrons and ions/neutrals is so large that the efficiency of energy transfer by elastic collisions can be neglected. On the contrary, many inelastic collisions can occur when a neutral species collides with an electron. These inelastic collisions can be divided into three main categories: excitation, dissociation and ionization. During excitation, the neutral species reaches an energy level higher than its fundamental level. When the excited species returns to the fundamental level, it emits a photon with an energy equal to the difference of energy between the excited state and the fundamental state. When a collision leads to dissociation, the molecular neutral species splits into two different neutral species. These species therefore have an unsatisfied bond and are a highly reactive species. Finally, an ionization process leads to the formation of a positive ion and a secondary electron. This last mechanism is responsible for sustaining a large enough electron density in the plasma. Some inelastic collisions can lead to multiple processes like dissociative ionization, and additional processes (not reported here) can lead to negative ion formation (e.g. electron attachment). The probability of a process occurring depends on the electron density, neutral density, the electron temperature and the cross-section of the reaction. Typical cross-sections for Cl_2/e^- collisions are represented in Figure 2.1 [CHE 12]. From this figure, it is clearly understandable that dissociation reactions are much more likely to happen than ionization. Therefore, the density of the charged species is much lower than the density of dissociation products. This trend can be generalized to other plasmas. From this figure, it is also understandable that a neutral species that is in minority as a neutral in the plasma may lead to a dominant ion if the ionization cross-section has a low threshold energy.

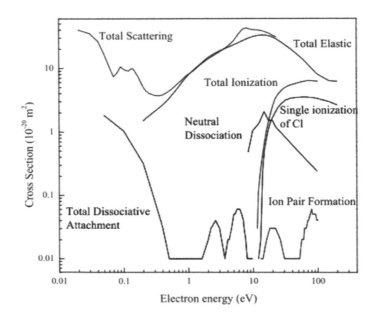

Figure 2.1. *Typical EEDF and cross-sections of Cl_2 dissociation and ionization. From Chen and Chang [CHE 12]*

   The consequence of the various inelastic collisions between the electrons and neutral species is the formation of a large variety of species in the plasma including electrons, ions, molecules, radicals and photons. The neutral species can originate from the injected gas molecules, and also from etch products or materials from the plasma chamber walls. Figure 2.2 illustrates the typical species in a SF_6-based plasma used for silicon etching. Upon electron impact, SF_6 molecules are dissociated/ionized, feeding the plasma with SF_x molecules, F radicals and SF_x^+ and F^+ ions. Electron (dissociative) attachment can also form negative ions such as SF_5^-. The excited species emit light with a specific wavelength when they go back to the fundamental state. The emitted light goes from the infrared to the vacuum ultra violet. Silicon reaction with fluorine provides SiF_2 and SiF_4

etch by-products that are eventually dissociated/ionized in the plasma. The ions that are accelerated through the sheath impact the chamber walls that are fluorinated by the plasma and can be sputtered, feeding the plasma with Al-containing species. This list is non-exhaustive, but gives a broad idea of the variety of species that are available in a typical etching plasma.

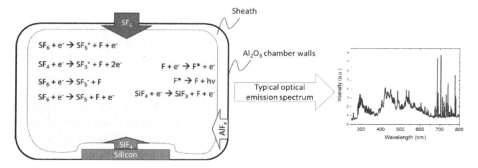

Figure 2.2. *Schematic of a SF$_6$-based plasma used for silicon etching [PIC 86] and typical optical emission spectrum of such plasma*

To sustain the plasma, an external electron excitation is required. For plasma etching applications, this excitation is generally performed using an external radio frequency electromagnetic field that accelerates the electrons. The frequency is generally too large for the low mobility ions to follow the electromagnetic field variations, while the light and mobile electrons can follow the instantaneous variations of the electromagnetic field. Therefore, most of the power is delivered to the electrons. There are several modes for RF excitation: capacitive coupling, inductive coupling and microwave coupling. In addition to the plasma generation power supply, additional power may be provided to the substrate to increase the sheath potential in front of the substrate and control the ion energy up to several hundreds of eV. Therefore, the ions that impinge the substrate during plasma etching can be much more energetic than 15 eV. To ensure a proper injection of power, a matching unit also called the match box is used to match the output impedance of the power supply and the impedance of the plasma.

2.1.2. *Typical plasma etch tools in the semiconductor industry*

The semiconductor industry uses plasma etching extensively for integrated circuit fabrication. Current technologies use 300 mm silicon wafers, and 450 mm wafers are considered for future generations. Therefore, the plasma etch tools must handle 300 mm wafers and provide excellent plasma uniformity across the whole wafer. There are a few companies currently able to provide and support such tools. The major companies are LAM RC, Tokyo Electron Limited, Applied Materials and Hitachi. In 2012, plasma etch tools corresponded to a market of ~6 billion of US dollars [QIU 13].

A typical plasma etch tool is divided into several parts as illustrated in Figure 2.3. Wafers are provided in Front Opening Unified Pods (FOUP) that are placed on load ports. The FOUP opening is robotized and a robotized arm picks the wafer up in the factory interface (FI) where it can eventually be aligned. Load locks are then used to transfer the wafers from atmospheric pressure in the FI to vacuum (typically few hundreds of mTorr) in the transfer chamber. Several plasma etch process modules (either identical or different chambers) can be mounted on the same vacuum transfer chamber. Etch chambers operate the process on single wafers. The wafer is electrostatically clamped on the chuck with backside helium flux to control the temperature.

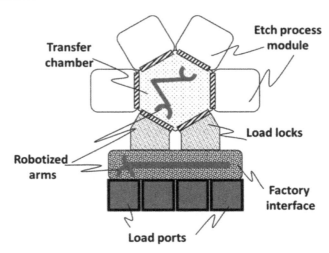

Figure 2.3. *Schematic view of a typical plasma etch tool*

Various plasma excitation modes are currently used in the semiconductor industry for plasma etching. Capacitive plasmas (CCP for capacitively coupled plasma) use two electrodes to provide the energy to the plasma, as illustrated in Figure 2.4(a). The substrate sits on one electrode. The capacitive coupling leads to a self-bias voltage on the electrodes, which also accelerates the ions and repels the electrons in the plasma. The time variation of the sheath voltage leads to electron heating. With capacitive coupling, the plasma generation and the ion acceleration cannot be independently controlled.

Depending on the RF frequency, the efficiency of the power coupling with the ions and the electrons varies, as shown in Figure 2.4(b) [PER 05]. At low frequency, the power is highly coupled to the ions, which leads to very large ion energy and low plasma density. On the contrary, high frequency excitation mostly leads to electron heating and therefore to higher plasma density and lower ion energy. This phenomenon can be advantageously used with multiple frequency plasma etch tools where the plasma is excited with two power supplies: a "low frequency" (typically few MHz) power supply that controls the ions energy and a "high frequency" (more than 20 MHz) power supply that controls the plasma density and therefore the ion flux and plasma dissociation. The latest generation of CCP even use a third frequency to give larger versatility to the plasma etch tools.

One variation of CCP tools is the magnetron enhanced plasmas, where a magnetic field surrounds the plasma chamber to confine the electrons in the plasma. To maintain good plasma uniformity, the external magnetic field is rotated either by rotating the physical magnets or by off-phase excitation of the surrounding electromagnets.

Another variation of the CCP tools is the DC superposition that consists of polarizing the top electrode with a DC potential (typically around 1,000 V). The top electrode must therefore be conducting and is typically made of silicon. The DC superposition accelerates the ions towards the top electrode, which generates secondary electrons. These secondary electrons are repelled to the plasma and accelerated by the DC potential. This creates a beam of very high energy electrons that can cross the plasma without any collision and reach the substrate on the opposite electrode. The beam of ballistic electrons is anisotropic and helps the process by improving the photoresist etch resistance and reducing the differential charges in the etched patterns [XU 08, ZHA 10].

With current CCP etch tools, the plasma density ranges between 10^8 and 10^{11} cm^{-3}. The plasma pressure ranges between a few tens of mTorr and a few hundreds of mTorr. Ion energy can reach more than 1,000 eV. Such tools are mostly used for dielectric etchings and interconnect fabrication.

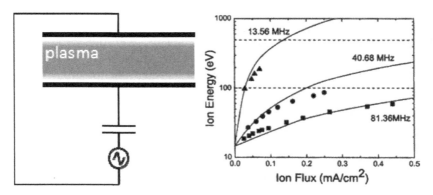

Figure 2.4. *(left). Schematic view of a CCP etch tool; (right). Ion energy versus ion flux for various excitation frequency in a 15 mTAr plasma. From Perret et al. [PER 05]*

The second family of plasma etch chambers uses inductive coupling to provide energy to the plasma. In these conditions, an external coil is excited with RF current, which generates an electromagnetic field in the plasma. This electromagnetic field accelerates the electrons in loops under the antenna. With such a configuration, plasmas can operate at lower pressures, and the plasma density is much larger than with capacitive coupling. An additional power supply is capacitively coupled to the plasma to control the sheath voltage and the ion energy. Therefore, the plasma generation is decoupled from the ion acceleration and the operator can independently control the plasma density (with the coil power) and the ion energy (with the capacitive power). Typical chamber configurations are shown in Figure 2.5.

Various configurations of the coil antenna have been designed by tool suppliers. When the antenna is a planar spiral, the configuration is called transformer coupled plasma. Another configuration uses a vertical coil on the top of the chamber wall as antenna. In this case, the grounded end of the antenna is placed on the plasma chamber and the high potential end is several centimeters above the plasma chamber to reduce capacitive coupling. With such configurations, multiple antenna are required to keep a uniform plasma. One last configuration uses a coil around the chamber walls and the plasma is, in this case, localized inside the coil. These three chamber configurations are illustrated in Figure 2.5.

With the current ICP etch tools, the plasma density ranges from 10^9 to 10^{12} cm^{-3}. The plasma pressure ranges between a few mTorr and a few tens of mTorr. Ion energy can reach several hundreds of eV. Such tools are mostly used for silicon and metals etching.

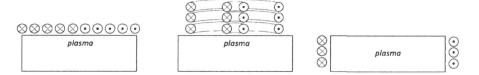

Figure 2.5. *Schematic of various configurations of inductive plasma chambers*

Recently, pulsed ICP plasmas have been introduced in the industry. The hardware of the pulsed ICP etch tool is very similar to standard ICP etch tools, except for the capability of pulsing the source and/or bias power supplies [BAN 12]. Several plasma pulsing modes illustrated in Figure 2.6 can be used and the most standard ones are presented here:

– Bias pulsing only: in this case, the plasma source is not pulsed (CW for continuous wave) and the bias power supply is pulsed. This leads to little modification of the plasma itself, and gives the ion energy distribution function (IEDF) a bimodal shape with a peak at low energy from the OFF time (similar to a plasma without bias) and a peak at high energy from the ON time (similar to a CW bias power).

– Synchronized pulsing: in this case, the source and the bias power are both pulsed. In standard synchronized pulsing, the source and the bias power supplies would be turned ON and OFF simultaneously. Some variants may include a delay and different duration of the ON time between the source and the bias power supplies. Pulsing the source power strongly changes the plasma and, in particular, its dissociation and density. Lower plasma densities and dissociation are observed when the plasma is pulsed, and the density and dissociation decrease when the duty cycle decreases. This results in a lower average ion flux with lower average ion energy. During the plasma OFF time, the sheath may collapse, leading to a flux of ions with a very low energy. Negative charges can also eventually reach the surfaces. This results in an average IEDF with a peak at very low energy (~1 eV) from the OFF time and a peak at high energy (from the ON time). However, there is no correspondence between the IEDF of pulsed plasmas and the CW counterpart. The presence of low energy ions and negative charges during the OFF time is useful for reducing the differential charges during plasma etching (see section 2.3.1) [BAN 12].

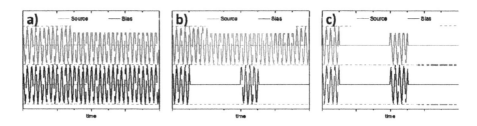

Figure 2.6. *Illustration of the main inductive plasma operation modes: a) continuous wave (CW); b) pulsed bias; c) synchronous plasma pulsing*

The last family of plasma etch chambers uses microwaves to excite the electrons close to their resonance frequency. Typical chamber configurations are presented in Figure 2.7. In electron cyclotron resonance (ECR) plasmas, an external magnetic field is used to make electrons rotate around the magnetic field lines at a frequency called the cyclotron frequency. A microwave is injected to the plasma. The magnetic field is chosen so that the cyclotron frequency matches the microwave frequency. Therefore, the electrons are excited by a resonance phenomenon. The power is solely coupled to the electrons and high plasma densities can be obtained with ECR plasmas. Another example of microwave plasmas is the radial line slot antenna (RLSA), where the surface waves on a specific electrode generate a localized high density plasma under the electrode. In both cases, an additional capacitively coupled power supply is used to accelerate the ions towards the substrate. Microwave plasmas can operate at very low plasma pressure (few mTorr) and lead to high density plasmas (10^{11}–10^{13} cm^{-3}). Their drawbacks are the difficulty to match the plasma and to obtain a uniform plasma.

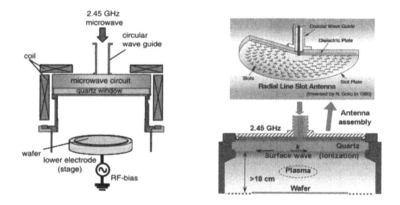

Figure 2.7. *Illustration of microwave-based plasma chambers: (left) electron cyclotron resonance chamber (adapted from Maeda et al. [MAE 12]) and (right) radial line slot antenna (from Zhao et al. [ZHA 13])*

Table 2.1 presents the major characteristics of the different modes of plasma excitations as well as their typical applications in the semiconductor industry for plasma etching.

Mode	Variant	Pressure [mTorr]	Density [cm^{-3}]	Ion Energy [eV]	Application
CCP	Simple frequency	10's to 100's	10^9-10^{10}	100's to 1000's	Interconnects fabrication Contacts etching
	Multi frequency	10's to 1000's	10^9-10^{11}	10's to 1000's	Interconnects fabrication Contacts etching
	Magnetically enhanced	10's to 1000's	10^9-10^{11}	100's to 1000's	Interconnects fabrication Contacts etching
	DC superposition	10's to 1000's	10^9-10^{11}	100's to 1000's	Interconnects fabrication Contacts etching
ICP	Continuous Wave	1's to 10's	10^{10}-10^{12}	15 to 1000's	Metal, Silicon etching
	Bias pulsing	1's to 10's	10^{10}-10^{12}	15 to 1000's	Metal, Silicon etching High aspect ratio patterns etching
	Synchronized pulsing	1's to 10's	10^{10}-10^{12}	1's to 1000's	Metal, silicon etching High aspect ratio patterns etching Stopping on thin layers
Micro Wave	Electron Cyclotron Resonance	1 to 10	10^{10}-10^{12}	15 to 100	Metal, silicon etching
	Radial Line Slot Antenna	1's to 100	10^{11}-10^{12}	10 to 100	Metal, silicon etching. Low damage processes

Table 2.1. *Major characteristics of the different modes of plasma*

2.2. Plasma surface interactions during plasma etching

To etch a material, it is necessary to remove the atoms from a substrate. This removal can be purely physical with a transfer of momentum from an incident species to the atom of the surface that can eventually leave the surface. This physical effect is used in sand blasting where the physical impact of sand beads erodes a surface. The removal can also be purely chemical with the reaction of the atoms of the surface with a chemical agent that forms reaction by-products that can eventually be washed out of the surface. In a plasma, ions with large energy impinge on the surface, and can therefore be used to physically etch a material. On the other hand, reactive radicals are formed in the plasma and can react with the material. If the reaction product is volatile at surface temperature and plasma pressure, it can desorb from the surface, which leads to the removal of the material. Therefore, both physical and chemical phenomena take place.

The physical and the chemical etching can also work together and help each other in removing the material. This synergetic effect is very efficient in plasma etching and has been evidenced by Coburn and Winters in 1979, as shown in Figure 2.8 [COB 79]. In their experiment, Coburn and Winters exposed a silicon sample to XeF_2 gaz. In this case, only chemical reactions can explain silicon etching and a slow etch rate of less than 1 nm/min is observed. They then added a flux of 450 eVAr^+ ions towards the surface. The etch rate rises to more than 6 nm/min. Finally, they turned off the XeF_2 gas flow and kept only the ion beam on. The etch rate drops back to less than 1 nm/min in a purely physical etching. This experiment shows that a synergetic effect between the physical impact (ion bombardment) and chemical reactions (XeF_2 gas) leads to a strong increase in the etch rate compared to each single phenomenon. This kind of experiment has also been reproduced for other gases and ions [COB 79].

We will now discuss in more details the phenomena responsible for physical and chemical etching, and for the synergetic effect between physical and chemical etching by looking at the etching systems that use radicals only or ion beams instead of a plasma. Such systems do not meet the requirements of advanced integrated circuit fabrication, but they illustrate the phenomena involved during plasma etching. The actual plasma etching mechanisms are then discussed. Finally, we will discuss the impact of the materials exposed to the plasma on the plasma etching processes.

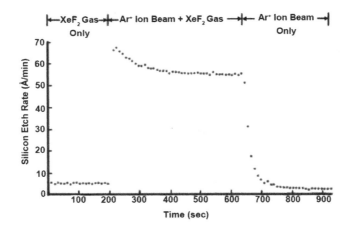

Figure 2.8. *Illustration of the synergetic effect between reactive reaction and ion bombardment. From Coburn and Winters [COB 79]*

2.2.1. *Purely spontaneous etching (downstream plasmas)*

Downstream plasmas (also called remote plasmas) are systems where the plasma chamber is deported from the substrate chamber. Neutral species can diffuse from the plasma chamber to the substrate chamber but ions are repelled or neutralized before they reach the substrate chamber. In such a system, only the radicals of the plasma are available to etch the material. The etching mechanism can be divided into three steps, as illustrated in Figure 2.9 for the case of silicon etching with chlorine. First, the reactive neutrals reach the substrate and adsorb at the surface. Second, a chemical reaction occurs between the reactive neutral and the atoms from the surface. Finally, the reaction by-product desorbs from the surface. For such reactions to occur, it is necessary for the products of the reaction to be volatile at the process conditions.

Each of the three steps can limit the etching. At very low partial pressure of the reactive species or when sticking species block adsorption sites, the first step may be limiting the etching. When the reaction is slow or when there is an energy barrier in the reaction, the second step can be limiting. Finally, if the reaction product has a low volatility, the third step is limiting the etching.

This purely chemical etching (also called spontaneous etching) is very similar to the etching processes that occur in wet chemical baths. The incoming species are isotropic, which leads to an isotropic plasma etching.

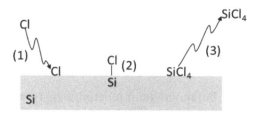

Figure 2.9. *Illustration of the three steps of chemical etching: (1) radicals adsorption; (2) reaction; (3) by-product desorption*

2.2.2. *Purely physical etching (ion beam etching)*

Ion beam etching (IBE) consists of using a beam of high energy ions from a neutral gas (typically argon) to sputter away atoms from a target. Such systems are used for surface analysis (e.g. secondary ion mass spectrometry, focused ion beams, etc.). The concept of physical etching (also called sputtering) relies on a transfer of momentum from an incoming ion towards the atoms of the target, as illustrated in Figure 2.10. If the momentum is reversed, the atoms from the target would leave the target, leading to etching. When the incoming ion collides with the atoms of the target, elastic collisions lead to a transfer of momentum, which is very efficient when the mass of the ion and the atom are close. The displaced atom can eventually collide with other atoms, which results in collision cascades. One or several atoms of the surface can be ejected in the process. The etching yield is defined as the ratio between the number of incoming ions and ejected atoms from the target. This phenomenon being purely physical, the etch yield of two different materials (e.g. Si and SiO_2) would be very close, which makes this process less selective.

The momentum of a particle is defined as the product of its mass and its velocity. The larger the incoming ion's momentum is, the greater the chance that atoms from the target will be sputtered away. Therefore, the etch yield depends on the velocity that varies linearly with the square root of energy. In addition, the bonds between the target atoms must be broken, and the surface potential must eventually be overcome for the etching. As a result, physical etching also involves a threshold energy below which there is no sputtering. Finally, the yield depends on the atoms involved in the collision. The sputtering yield can therefore be expressed as $Y(E_i) = A.\left(\sqrt{E_i} - \sqrt{E_{th}}\right)$, where Y is the etch yield, E_i the ion energy and E_{th} the threshold energy [STE 89].

In physical etching, the momentum must also be reversed. When the ion reaches the target at a slightly off normal angle, a full reversal of the momentum is not necessary, which helps in the physical etching, as shown in Figure 2.10. The effect becomes more important when the incidence angle increases (with respect to the normal) until the angle is large enough for reflections to dominate. This angle is called the critical angle. In the latter case, the etch yield drops when the incidence angle further increases and the incident ion is reflected by the target without much momentum transfer. This dependence of the etch yield is illustrated in Figure 2.10. When a pattern is present on the target, the edges can be viewed as a continuum of the surface, from horizontal to vertical. Since the etch yield is maximum for a critical angle, the physical etching will select this angle and transfer it preferentially on the surface. This results in tapered patterns.

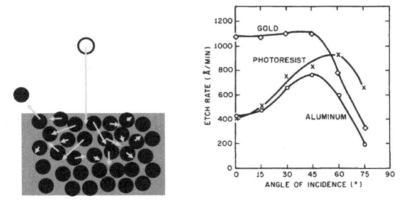

Figure 2.10. *(left) Illustration of collision cascades during sputtering. The incident atom is white and the arrow represents atom displacement. (right) Sputtering yield of gold, photoresist and aluminum as a function of the ion incident angle. From Manos et al. [MAN 89]*

It is noteworthy that when an ion reaches the surface, it is neutralized by a tunneling electron emitted by the surface [LIE 05]. Therefore, the species reaching the surface are hyper thermal neutrals with an incident energy identical to that of the incident ion. When the ion is molecular, the neutralization also results in a dissociation. In this case, fragments of the ion reach the surface with the same velocity as the incident ion, i.e. the energy is split between the constituents with the ratio of their mass over the mass of the ion [LIE 05].

With IBE, an etch yield close to 1 is generally obtained for an ion energy close to 1 keV. For silicon, this results in a slow etch rate of close to 1 nm/min for a current density of 1 mA.cm^{-2}. The etching only occurs in the direction of the ion flux. In a plasma, the ion flux is perpendicular to the wafer surface because of the plasma sheath. Therefore, ion-induced sputtering is an anisotropic process.

2.2.3. Reactive IBE (RIBE)

Reactive IBE (RIBE) is identical to IBE, except for the nature of the ions. For IBE, ions come from rare gases and are not reactive. For RIBE, ions come from reactive elements (in particular halogens) that can react with the substrate. Figure 2.11 shows etch yield comparison for silicon etched in IBE (with Ar$^+$) and in RIBE mode (with Cl$^+$). The rare gas ion for IBE is Ar$^+$ for comparison with atoms of similar masses (35.45 vs. 39.95 amu). We can see that both RIBE and IBE present a square root evolution of the etch yield with the ion energy, as explained for IBE. However, the etch yield is much larger for RIBE mode than for IBE mode, and the threshold energy is lower for RIBE than for IBE. This shows that adding some chemical reactivity to IBE strongly enhances the etching. The reactive species provided as reactive ions can form a mixed surface between the reactive species and atoms from the substrate (e.g. a mixed Si/Cl surface) that is easier to sputter away than the substrate itself. The ion energy will favor the formation of a several nanometer-thick reactive layer, in which the reactive species are implanted and mixed.

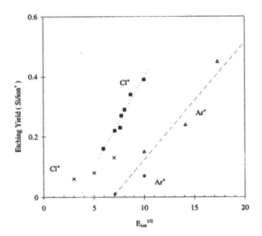

Figure 2.11. *Sputtering yield of polysilicon using Ar+ and Cl+ ion beams. From Chang et al. [CHA 97]*

2.2.4. *Chemically assisted IBE (CAIBE)*

Chemically assisted IBE (CAIBE) uses an ion beam to etch a material in a chamber with background gas. In this case, the ions are not reactive (typically Ar^+ ions), but the background gas provides some chemical reactivity to the process. Figure 2.12 shows some published results of the silicon etch rate in CAIBE systems using either Cl_2 or SF_6 as a background gas. We can see in these figures that the etch yield is strongly enhanced when a reactive gas is introduced in the chamber. When the flow of background gas increases, the etch yield tends to increase. In addition, increasing the ion energy still increases the etch yield. This shows the combined effect of the chemical reactions between the substrate and the background gas and the physical impact of the ion bombardment.

Molecules from the background gas can adsorb on the surface of the substrate and eventually react with the substrate to form volatile species. The ion-induced surface damage provides adsorption sites at the surface of the substrate, favoring the first step of the chemical etching. It also provides energy to the surface, which favors the second step of the chemical etching, eventually providing enough energy to overcome the energy barrier of the reaction. Finally, the ion bombardment can sputter away low volatility reaction products, helping the third step of the reaction.

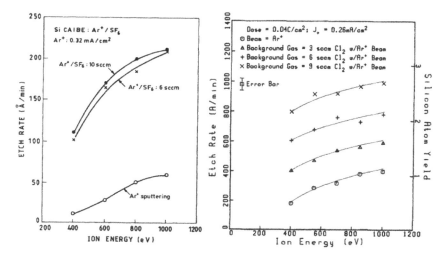

Figure 2.12. *Etch rate of silicon in CAIBE using (left) SF$_6$ (from Ray et al. [RAY 95]) and (right) Cl$_2$ (from Chinn et al. [CHI 84]) as a background gas.*

2.2.5. *Real plasma etching mechanisms*

The various beam etching technologies presented above have shown the synergetic effects between the chemical and the physical components of etching. In real plasma etching, the substrate receives a flux of various species. Reactive neutrals from the plasma directly participate in the etching by forming volatile etch by-products. Ion bombardment provides some assistance to the three steps of the chemical etching, as explained for CAIBE. In addition, the ions can also provide reactive species that participate in the etching and are eventually implanted in the reactive layer, as explained for RIBE. Photons from the plasma are also absorbed in the upper most surface. Their impact on the etching process is not clearly understood yet, but they can provide energy and favor the reaction and the desorption at the surface [SHI 12].

The synergetic effect between the ions and radicals is generally the dominant mechanism during plasma etching. It provides a large etch rate in the direction of the ion bombardment that is perpendicular to the surface because of the plasma sheath. Therefore, the etching is strongly favored in the direction normal to the surface compared to the in-plane direction. This anisotropic character of plasma etching is the major advantage of plasma etching compared to wet cleaning solutions. Purely physical and purely chemical effects play a second order role in the etching, and lead to process imperfections such as profile distortion, materials damage and mask sputtering.

We have seen in section 2.1.1 that the plasma contains much more neural species than ions. Depending on the plasma density, dissociation and pressure, the ratio between the radicals flux and the ion flux to the surface may vary, which will change the etching regime. This is illustrated in Figure 2.13 using the beam experiment. When the radicals to ions ratio is low, the etch rate is mostly limited by the radicals' flux. Therefore, small variations in the radical's flux lead to large variations in the etch yield. On the contrary, the variations in the ion energy have less impact on the etch yield. When the radicals to ions ratio is large, the etch rate is mostly limited by the ions and the etch yield hardly varies when the radicals' flux changes. On the contrary, changing the ions' energy has a large impact on the etch yield. These two regimes are called the "neutral limited regime" and the "ion limited regime". This was evidenced using beam experiments, but similar mechanisms also occur during plasma etching.

In real plasma etching, the plasma also contains molecules and their products of dissociation that may participate in the etching by favoring or disfavoring the etching. Stable molecules will have a low impact on the etching since they are much less reactive than radicals. They can nevertheless provide reactive species to the surface, especially when their adsorption is dissociative. Sticking molecule also tend to form a thick deposit at the surface of the substrate. Such a deposit blocks the adsorption of reactive species and protects the substrate from the ion bombardment. Therefore, they tend to block the etching and are referred to as etch inhibitors. If their composition contains reactive species (e.g. fluorine in fluorocarbon species), they can also provide a feedstock of reactive species that participate in the etching, provided the deposited thickness remains below the penetration depth of the ions in the layer.

Etch inhibitors can be helpful in controlling the etch profile. Indeed, their deposition can be avoided at the etch front thanks to the high energy ion bombardment, while their deposition will be possible at the pattern sidewalls where the ions cannot sputter them away. In this case, they are called "passivation layers" and help in preventing isotropic etching by inhibiting the spontaneous reactions between the substrate and the radicals from the plasma, and by absorbing the energy of the scarce ions reaching the sidewalls.

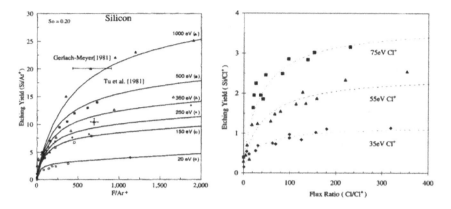

Figure 2.13. *Silicon etch yield as a function of the radicals to ions flux ratio. (left) For F and Ar⁺. From Gray et al. [GRA 93]. (right) For Cl and Cl⁺. From Chang et al. [CHA 97]*

2.2.6. *Impact of plasma parameters on plasma etching*

All parameters of the plasma impact the interactions between the plasma and the surfaces. To control an etch process, the operator must choose the right knobs to set up the plasma parameters. In turn, the plasma parameters define the outcome of the etching process. However, the exact impact of the process knobs on the plasma parameters is hardly known. Similarly, the exact role of the plasma parameters on the etch results is also not fully understood. This makes the empirical process development using trial and error or using design of orthogonal experiments (DOE) the dominant approach to plasma process development nowadays.

In general, increasing the source power produces a more dissociated plasma with a higher density and ion flux. Increasing the bias power increases the ion energy, which favors the process anisotropy but decreases the selectivity and leads to mask erosion. Changing the pressure has an impact on the electron temperature and neutrals density, but it is not straightforward to anticipate whether the radical flux will increase or not since the dissociation and plasma density change when the pressure changes. At a higher pressure (several tens of mTorr), the sheath becomes collisional, which increases the ion angular distribution function and reduces the average ion energy, resulting in some lateral etching.

2.2.7. Reactor walls and plasma etching

The plasma's physical properties are also impacted by the chamber walls. Indeed, most of the chemical reactions such as recombination occur at the chamber walls. Therefore, changing the chamber walls' material and/or temperature has a direct impact on the plasma properties. This phenomenon is illustrated here in the case of a Cl_2 plasma [CUN 07]. Figure 2.14 shows the percentage of chlorine radicals in a Cl_2 plasma as a function of the plasma source power for various chamber walls' materials. For SiOCl-coated chamber walls, the recombination coefficient is very low (estimated at 0.05), which means that the Cl radicals hardly recombine on the chamber walls to form Cl_2. This results in a highly dissociated plasma (almost 100% of Cl is present as radical in the plasma) even at a low source power. On the contrary, Cl radicals have more chance to recombine on chamber walls with a larger recombination coefficient (e.g. 0.3 for AlF chamber walls), which results in a lower plasma dissociation even at a large source power. These variations of the radicals' density directly impact the plasma etch process, especially when the process is in the radical limited regime.

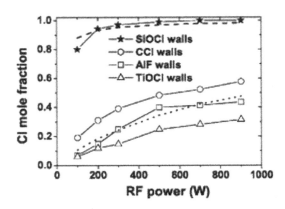

Figure 2.14. *Percentage of Cl atoms as radicals in a Cl_2 ICP plasma as a function of the source power. From Cunge et al. [CUN 07]*

On the contrary, the plasma etch process also impacts the chamber walls. The ion bombardment can sputter away particles from the chamber walls that could contaminate the etched wafers. The coating from the chamber walls could also be etched by the plasma etch process. Finally, etch by-products or etch inhibitors from the plasma can deposit on the chamber walls and build up a relatively thick layer. Therefore, the chamber wall conditions evolve during the etch process, which in turn changes the etch process. They can also induce contamination when they redeposit on another wafer that is etched subsequently. These phenomena lead to memory effect and first wafer effects, where the outcome of an etch process depends on the chamber history. To obtain reproducible process conditions, it is possible to use cleaning processes to recover identical chamber conditions (referred to as clean chamber walls) before each of the processes. Typical cleaning processes use a SF_6/O_2-based plasma for silicon etch chambers or O_2-based plasmas when the etch process uses fluorocarbon plasmas [CHE 07]. However, a specific cleaning process should be developed depending on the materials that have been etched in the chamber [RAM 07]. In addition, a seasoning step is often performed before etching a wafer or a batch of wafers either by exposing a dummy wafer to the etch process or by using a depositing plasma to coat the chamber walls with thin and controlled materials. In the latter case, $SiCl_4/O_2$-based plasmas are the standard processes.

2.2.8. *Plasma changes due to etching*

The surface of the etched wafer can also impact the plasma parameters. Indeed, etching processes consume radicals from the plasma to form etch by-products that desorb from the wafer and feed the plasma. This results in a depletion in etching radicals and the presence of etch by-products and their dissociation products when a material is etched. The depletion in etch radicals results in a decrease of the etch rate, especially when the etching operates in a neutral limited regime. This phenomenon is called a loading effect. The etch by-products and their products of dissociation can play the role of etch inhibitors that would also slow down the etching. An excellent example of the impact of plasma etching on the plasma parameters was presented by Cunge et al., in the case of silicon etching in a $HBr/Cl_2/O_2$ plasma [CUN 02]. As shown in Figure 2.15, the flux of the halogen-containing ions decreases when the bias power (and thus the etch rate) increases. On the contrary, the flux of the silicon-containing ions strongly increases with the increase in the bias power. With their conditions, Cunge *et al.* report that up to 50% of the ions that reach the silicon surface already contain silicon atoms.

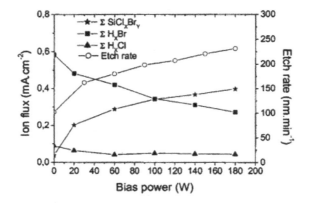

Figure 2.15. *Ion flux composition and etch rate of silicon in a HBr/Cl₂/O₂ plasma during silicon etching as a function of the bias power. From Cunge et al. [CUN 02]*

2.3. Patterns transfer by plasma etching

Plasma etching processes are used to transfer the patterns defined in a mask towards another material. The specificity of plasma etching is the anisotropy of the process that enables a conservation of the dimension of the mask into the etched patterns and the fabrication of vertical patterns. In the following, we will discuss how passivation layers control the etched profile and how the etch process is impacted by the patterns themselves. We will then discuss the typical pattern profiles obtained by plasma etching.

2.3.1. *Pattern-related phenomena*

We have shown in section 2.2.5 that etch inhibitors can deposit on the pattern sidewalls and form passivation layers. These passivation layers are extremely important in the plasma etching processes since they control the final profile of the patterns. The passivation layers are necessary to prevent lateral etching of the patterns. Even if the synergy effect between radicals and ions favors the etching in the direction of the ion flux, the high density of radicals in the plasma can still lead to some spontaneous etching at the pattern sidewalls. In addition, the ion bombardment is not perfectly vertical and some ions can still reach the sidewalls. This is especially true when differential charges deflect the ions towards the sidewalls, as we will discuss later.

However, the deposition of passivation layers at the pattern sidewalls can also lead to pattern distortion. They tend to make the patterns larger along the process, which results in tapered profiles if the passivation layers are too thick. If their deposition rate is not uniform throughout the wafer, it would result in variations of the pattern profiles across the wafer.

There are two main geographical sources for the passivation layers' precursors, as shown in Figure 2.16. These precursors can come from the plasma and stick on the pattern sidewalls. In this case, the passivation layers tend to accumulate on the sidewalls, and the upper part of the sidewall (that is exposed longer to the plasma process) has a thicker passivation than the bottom part of the sidewall. In addition, patterns with different densities would have different passivation layers' thicknesses, since the neighboring pattern could reduce the flux of etch inhibitors to the sidewalls (as explained later for the aspect ratio-dependent etching). The second origin of the etch inhibitors could be the etch front. In this case, the low volatility species would be sputtered from the etch front by the high energy ion bombardment and get deposited on the sidewalls in the line of sight. Hübner *et al.* [HUB 92] modeled the deposition rate for both origins and have shown that the passivation layers are thinner and more uniform when they are deposited from the etch front. The sidewall passivation layer profile computed by Hübner is shown in Figure 2.16.

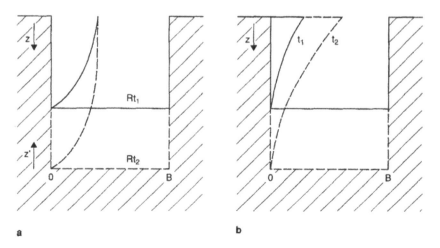

Figure 2.16. *Illustration of the mechanisms of passivation layer deposition and passivation layer profile computed by Hübner. (a) Redeposition of etch by-products; (b) plasma deposition. From Hubner et al. [HUB 92]*

The chemical source for the precursors of the passivation layer is also diverse. Dissociation products of molecules injected in the plasma can directly form passivation layers. Etch by-products from the etched material of the mask can also be precursors of the passivation layers. Finally, the deposition can result from the reaction of the etch by-products with dissociation products from molecules injected in the plasma. For example, passivation layers for silicon etching in $HBr/Cl_2/O_2$ plasma result from the oxidation of silicon-containing etch by-products on the sidewalls of the patterns by oxygen radicals [DES 01].

A second pattern-related effect in plasma etching is the building up of differential charges across the patterns. The quasi-neutrality of the plasma imposes an identical average flux of positive charges and negative charges during the process. The plasma sheath imposes an anisotropic flux of positive ions, while electrons that can follow the instantaneous variations of the sheath voltage reach the surface in bursts with an almost isotropic direction [ARN 91]. When the substrate presents a pattern, the sidewalls of the pattern hardly receive any ions, while they will receive electrons that can travel in the off-normal direction. On the contrary, at the bottom of the patterns, the flux of ions is not significantly affected by shadowing since the ions arrive vertically while the flux of electrons is strongly reduced due to the masking effect of the electron flux by the trench sidewalls (also called a shading effect). This results locally in a different average flux of positive and negative charges. If the material exposed to the plasma is insulating, charges accumulate until they induce enough modification of the electric field for the negative and positive charges' flux to equate, or for the charges to evacuate through the conduction paths. This steady state is reached within a few hundreds of micro- to milliseconds [KIN 96]. An example of differential charges computed by Hwang and Giapis is shown in Figure 2.17 in the case of silicon lines on SiO_2 etched with a photoresist mask [HWA 97a]. We can see that surface potential is larger when the aspect ratio of the opening increases, and that ions are directed towards the interface between Si and SiO_2. The differential charges that build up in patterns distort the trajectory of the ions that ends up with distortions of pattern profiles.

The differential charges depend on many parameters like the materials' conductivity, the plasma density and the electron temperature. From a geometrical point of view, the major parameter impacting the differential charges is the aspect ratio. Differential charges are particularly important when the insulator materials are etched. For conductor etching, the different flux of negative and positive charges leads to current flows that can eventually deteriorate the device when the etched structures present a large side to bottom ratio (so-called antenna structures) [HWA 97b].

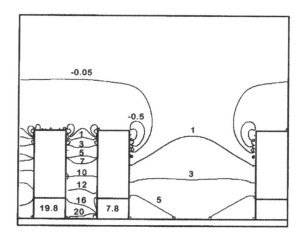

Figure 2.17. *Simulation of potential contour maps during silicon etching with a photoresist mask, stopping on a SiO$_2$ layer. The curves indicate equipotential lines and the number gives the value of the potential. From Hwang and Giapis [HWA 97a]*

We have seen that the aspect ratio is the dominant geometrical parameter for the differential charges' build up. We have also seen that the aspect ratio controls the thickness of the passivation layers when the precursors come from the plasma. The etch rate of trenches is also impacted by the aspect ratio of the patterns. This phenomenon is called ARDE (aspect ratio-dependent etching) or the aperture effect, pattern factor or RIE-lag. Figure 2.18 shows hole patterns with various diameters etched simultaneously in germanium. It is clear that the depth of the hole depends on its diameter, which shows that the average etch rate is lower when the diameter of the hole is smaller. The calculation of the instantaneous etch rate at different etch times reveals that the etch rate decreases when the etched depth increases, and that the etch rate is defined by the aspect ratio rather than by the diameter or the depth of the hole (see Figure 2.18(b)). The ARDE phenomenon is observed for many etch processes and in various materials. It is more severe for holes than for trenches. There are three main origins for this ARDE phenomenon, with the dominant one depending on the plasma process and/or the material being etched.

When the etch process is limited by the radicals' flux, small variations of the flux of radicals change the etch rate. When the radicals are very reactive, they would adsorb and react with the surface. Therefore, they would reach the bottom of the trenches only if they do not collide with the trench walls. As a result, the flux at the bottom of the trenches is reduced by the ratio between the solid angle of collection at the bottom of the trench and the solid angle of collection on a flat surface (2π). The solid angle at the bottom of a trench or a hole only depends on the pattern aspect ratio, as shown in Figure 2.18(b). The strong reduction of the solid angle – and thus the neutral flux – when the aspect ratio increases can explain the reduction of the etch rate when the aspect ratio increases.

When radicals are less reactive, they can bounce around in the trench/hole before reaching the bottom of the pattern or escaping from the pattern. This kind of transport is similar to a Knudsen transport in a slit or a pipe [COB 89]. The Knudsen transport occurs when the critical dimension of the pattern is smaller than the species' mean free path. In this case, the species are supposed to travel without collision from one wall to the other, and to escape from the wall with a cosine distribution. The transmission probability for a pipe can be calculated by Monte Carlo simulations and depends on the pipe aspect ratio [STE 86].

The last explanation of the ARDE is the formation of differential charges. The major geometrical factor impacting the differential charges is the aspect ratio. The net flux of ions at the bottom of the trenches is reduced when differential charges distort the ion direction. In addition, for insulator etching, positive charges built up at the etch front repel the ions, which further decreases the ion flux and the ion energy at the bottom of high aspect ratio patterns.

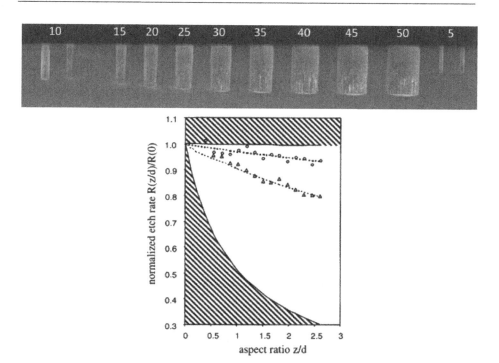

Figure 2.18. *(top) Circular holes etched in germanium with diameter varying from 5 to 50 μm. (bottom) Normalized etch rate of silicon in Cl₂-based plasma for two plasma conditions (circles and triangles). The hatched area corresponds to inaccessible regimes and are limited by the transport for a sticking coefficient of 0 (top) and 1 (bottom). From Lill et al. [LIL 01]*

Finally, we have seen that plasma etching has an impact on the plasma itself by consuming the etching radicals and feeding the plasma with the etch by-products (loading effect). This phenomenon can also be observed at the micrometer scale, when local variation in pattern densities can lead to local variations of the etch rate. The density of plasma etching radicals can be modified locally close to areas with a large surface to etch, compared to an area with a smaller surface to etch. This leads to local variations of the etch rate when the surface of material to etch changes. The so-called micro-loading phenomenon has been observed over distances up to 1 mm [HED 94]. It is reduced when the plasma pressure is reduced or when the gas flow rate increases.

2.3.2. *Pattern profile obtained by plasma etching*

The various pattern-related phenomena presented above also have an impact on the profile of the patterns. In general, the targeted pattern profile would be vertical, i.e. with the same width at the top and at the bottom of the pattern. However, the actual profile obtained may be different depending on the plasma etch conditions and the materials being etched. Figure 2.19 illustrates the typical pattern profiles that can be obtained after plasma etching. The naming of the various pattern profiles can depend on the application and from one research group/industrial to another.

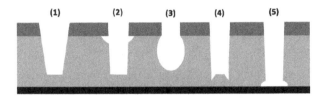

Figure 2.19. *Schematic illustration of pattern profiles obtained after plasma etching: (1) tapered profile, (2) undercut, (3) bowing, (4) microtrenching, (5) notching*

Tapered profiles present straight profiles with a linear variation of the width along the height of the pattern. Such patterns are usually observed when the passivation layers are deposited from the gas phase for isolated lines. In this case, there is a linear variation of the passivation layer thickness with the height of the pattern that linearly increases the apparent mask width. After the removal of passivation layers, the resulting pattern profile is tapered. A second origin for tapered profiles is the linear reduction of the mask width during the etch process (either by lateral etching of the mask or by faceting of the mask). In this case, the pattern width is reduced during the process, which results in tapered profiles. The last explanation for tapered profiles is the ions' reflection when they are reaching the sidewalls at a grazing angle. When the profiles are slightly tapered, the ions that reach the sidewalls reflect on the sidewalls rather than etching it, which prevents the straightening of the sidewalls by the vertical ion bombardment. A correct balance between passivation layer deposition and lateral etching can avoid tapered profiles.

Undercut profiles are defined as a lateral etching of the material under the mask. It originates from a lateral etching by plasma radicals. For an undercut profile to occur, the passivation layer under the mask must be removed. This removal can be due to ion bombardment when differential charges deflect the ions towards the sidewalls, or when the ions are reflected on the mask on the other side of the trench/hole pattern. The removal can also come from a chemical etching of the passivation layer by plasma radicals. Once the passivation layer is removed, plasma radicals can chemically etch the sidewalls, eventually with the assistance of the ions deflected by the differential charges or reflected from the mask on the other side of the pattern. This phenomenon is more easily observed when the passivation layer precursors originate from the etch front. In this case, there is no re-enforcement of the passivation layer along the process, as described by Hubner [HUB 92]. The undercut is generally only observed under the mask, where the flux of reactive species is largest (larger angle of collection) and where the passivation layer is more easily removed. However, in some cases, it can extend along the whole height of the patterns when the etched plasma can easily, spontaneously etch the material and the passivation layers, as is the case for amorphous carbon etching in oxygen-based plasmas. Undercut can be minimized by ensuring strong enough passivation layers to resist to the whole process.

Bowing profiles correspond to the arched sidewalls. It is particularly observed during trenches or holes etching in dielectrics. Bowing profiles are attributed to the lateral etching of the patterns by ions that are deflected by the differential charges. When insulators are etched, differential charges lead to a deflection of the ion flux towards the sidewalls. Bowed profiles are re-entrant profiles, i.e. the largest width of the trench/hole is not localized at the trench/hole entrance. This makes trench/hole filling very difficult, since the top of the pattern may close before the whole pattern is filled. To reduce bowing, thicker passivation layers may be deposited with the risk of getting tapered profiles. Increasing the bias power can also reduce the impact of the differential charges on the ions' direction. It is also possible to use technologies that reduce the differential charges (e.g. pulsed plasmas or DC superposition).

Microtrenching involves deeper etching close to the pattern sidewalls. It is explained by the reflection of the ions from the pattern walls that locally increases the total ion flux (direct ion flux plus reflected ion flux) at the bottom of the patterns. Depending on the pattern profiles, the reflected ion flux may be precisely localized at the edge between the wall and the etch front (focusing effect) leading to deep and narrow microtrenching, or may be reflected to a broader extent in the trench, leading to shallower and broader microtrenching [LAN 00]. Microtrenching also depends on the reflection mechanisms at the pattern walls. Specular reflection tends to lead to more focused reflected ions while broad reflection leads to broader microtrenches [VYV 00]. Since the microtrenching originates from a local variation of the ion flux, its observation indicates an ion flux limited regime. To reduce microtrenching, it is therefore possible to switch to regimes more limited by the radicals.

Notching is defined by a local lateral etching at the interface between the material being etched and the etch stop layer. This is a typical profile obtained during gate etching. Hwang and Giapis demonstrated that it originates from the deflection of the ion flux towards the edges when the etch process reaches the insulting bottom [HWA 97a]. In addition, the passivation layer may be thinnest close to the bottom of the pattern, especially when the precursors originate from the plasma. When the process reaches the etch stop layer, the reduction of the loading effect leads to an increase in the density of the reactive species that can attack the sidewalls where the passivation layer is the weakest. The notching extends during the etching time. To reduce notching, pulsed plasmas can be used to reduce the differential charges [SAM 94], or the processes can be used to strengthen the passivation layers during the over etch time.

2.4. Conclusion

Plasmas are complex media that contain many different species that are not perfectly known. The various species of the plasma interact with the materials exposed to the plasma such as the chamber walls and the substrate. The electric field created by the plasma sheath and eventually some additional bias power leads to the bombardment of the substrate by high energy ions. Gas molecule dissociation in the plasma creates etch radicals that can chemically react with the substrate. The combination of high energy ion bombardment and highly reactive radicals can be used to etch the substrate in an anisotropic way. Etch inhibitors that accumulate on the pattern sidewalls form passivation layers that prevent lateral etching and help in achieving anisotropic etching of patterns.

The presence of patterns on the wafer also impacts the etch process. Differential charges build up in the patterns and deflect the ions in non-vertical directions. Variation of the collection angle and Knudsen diffusion also change the radical flux, depending on the pattern aspect ratio. The etching of the material also changes the density of reactive species, both in the whole plasma and locally close to the etched area. These phenomena lead to various pattern profiles that differ from the targeted vertical profile.

All these plasma and pattern-related phenomena are not fully understood, and there is no straightforward way to anticipate the process outcome. Some solutions can be used to correct the pattern profiles, but they usually come at the expense of other pattern parameters. However, plasma etching processes have been able to provide satisfactory results so far, and no concurrent technology is available yet to replace the plasma etching processes for microelectronic device fabrication.

2.5. Bibliography

[ARN 91] ARNOLD J.C., SAWIN H.H., "Charging of pattern features during plasma etching". *Journal of Applied Physics*, vol. 70, no. 10, pp. 5314–5317, 1991.

[BAN 12] BANNA S., AGARWAL A., CUNGE G. *et al.*, "Pulsed high-density plasmas for advanced dry etching processes", *Journal of Vacuum Science & Technology A*, vol. 30, no. 4, p. 40801, 2012.

[CHA 97] CHANG J.P., SAWIN H.H., "Kinetic study of low energy ion-enhanced polysilicon etching using Cl, Cl_2, and Cl^+ beam scattering", *Journal of Vacuum Science & Technology A*, vol. 15, no. 3, pp. 610–615, 1997.

[CHE 07] CHEVOLLEAU T., DARNON M., DAVID T. *et al.*, "Analyses of chamber wall coatings during the patterning of ultralow-k materials with a metal hard mask: consequences on cleaning strategies", *Journal of Vacuum Science & Technology B*, vol. 25, no. 3, p. 886, 2007.

[CHE 12] CHEN F.F., CHANG J.P., *Lecture Notes on Principles of Plasma Processing*, Springer Science & Business Media, 2012.

[CHI 84] CHINN J.D., PHILLIPS W., ADESIDA I. *et al.*, "Ion beam etching of silicon, refractory metals, and refractory metal silicides using a chemistry assisted technique", *Journal of the Electrochemical Society*, vol. 131, no. 2, pp. 375–380, 1984.

[COB 79] COBURN J.W., WINTERS H.F., "Ion- and electron-assisted gas-surface chemistry – an important effect in plasma etching", *Journal of Applied Physics*, vol. 50, p. 3189, 1979.

[COB 89] COBURN J.W., WINTERS H.F., "Conductance considerations in the reactive ion etching of high aspect ratio features", *Applied Physics Letters*, vol. 55 no. 26, pp. 2730–2732, 1989.

[CUN 02] CUNGE G., INGLEBERT R.L., JOUBERT O. *et al.*, "Ion flux composition in $HBr/Cl_2/O_2$ and $HBr/Cl_2/O_2/CF_4$ chemistries during silicon etching in industrial high-density plasmas", *Journal of Vacuum Science & Technology B*, vol. 20, no. 5, pp. 2137–2148, 2002.

[CUN 07] CUNGE G., SADEGHI N., RAMOS R., "Influence of the reactor wall composition on radicals' densities and total pressure in Cl2 inductively coupled plasmas: I. Without silicon etching", *Journal of Applied Physics*, vol. 102, no. 9, p. 93304, 2007.

[DES 01] DESVOIVRES L., VALLIER L., JOUBERT O., "X-ray photoelectron spectroscopy investigation of sidewall passivation films formed during gate etch processes", *Journal of Vacuum Science & Technology B Microelectron Nanom Structure*, vol. 19, no. 2, p. 420, 2001.

[GRA 93] GRAY D.C., TEPERMEISTER I., SAWIN H.H., "Phenomenological modeling of ion-enhanced surface kinetics in fluorine-based plasma etching", *Journal of Vacuum Science & Technology B*, vol. 11, no. 4, pp. 1243–1257, 1993.

[HED 94] HEDLUND C., BLOM H.O., BERG S., "Microloading effect in reactive ion etching", *Journal of Vacuum Science & Technology A*, vol. 12, no. 4, pp. 1962–1965, 1994.

[HUB 92] HÜBNER H., "Calculations on deposition and redeposition in plasma etch processes", *Journal of the Electrochemical Society*, vol. 139, no. 11, pp. 3302–3309, 1992.

[HWA 97a] HWANG G.S., GIAPIS K.P., "On the origin of the notching effect during etching in uniform high density plasmas", *Journal of Vacuum Science & Technology B Microelectronics and Nanometer Structures*, vol. 15, no. 1, p. 70, 1997.

[HWA 97b] HWANG G.S., GIAPIS K.P., "On the origin of charging damage during etching of antenna structures", *Journal of the Electrochemical Society*, vol. 144, no. 10, p. L285, 1997.

[KIN 96] KINOSHITA T., HANE M., MCVITTIE J.P., "Notching as an example of charging in uniform high density plasmas", *Journal of Vacuum Science & Technology B Microelectronics and Nanometer Structures*, vol. 14, no. 1, p. 560, 1996.

[LAN 00] LANE J.M., KLEMENS F.P., BOGART K.H.A. *et al.*, "Feature evolution during plasma etching. II. Polycrystalline silicon etching", *Journal of Vacuum Science & Technology A Vacuum, Surfaces, Film*, vol. 18, no. 1, p. 188, 2000.

[LIE 05] LIEBERMAN M., LICHTENBERG A.J., *Principles of Plasma Discharges and Materials Processing*, 2nd ed., pp. 1–757, Wiley, 2005.

[LIL 01] LILL T., GRIMBERGEN M., MUI D., "*In situ* measurement of aspect ratio dependent etch rates of polysilicon in an inductively coupled fluorine plasma", *Journal of Vacuum Science & Technology B Microelectronics and Nanometer Structures*, vol. 19, no. 6, p. 2123, 2001.

[MAE 12] MAEDA K., OBAMA S., TAMURA H. *et al.*, "Study on the distribution control of etching rate and critical dimensions in microwave electron cyclotron resonance plasmas for the next generation 450mm wafer processing", *Japanese Journal of Applied Physics*, vol. 51, no. 8, pp. 1–6, 2012.

[MAN 89] MANOS D.M., FLAMM D.L. (Eds), *Plasma Etching: An Introduction*, Elsevier, 1989.

[PER 05] PERRET A., CHABERT P., JOLLY J. *et al.*, "Ion energy uniformity in high-frequency capacitive discharges", *Applied Physics Letters*, vol. 86, no. 2, p. 21501, 2005.

[PIC 86] PICARD A., TURBAN G., GROLLEAU B., "Plasma diagnostics of a SF_6 radiofrequency discharge used for the etching of silicon", *Journal of Physics D: Applied Physics*, vol. 19, no. 6, p. 991, 1986.

[QIU 13] QIU T., BEAUBOIS J., "Semiconductor equipment shrinkage brings growth, but not for everyone", *Berenberg Equity Research Report*, July 22nd, 2013.

[RAM 07] RAMOS R., CUNGE G., JOUBERT O. *et al.*, "Plasma/reactor walls interactions in advanced gate etching processes", *Thin Solid Films*, vol. 515, no. 12, pp. 4846–4852, 2007.

[RAY 95] RAY S.K., MAITI C.K., LAHIRI S.K., "Chemically assisted ion beam etching of silicon and silicon dioxide using SF_6", *Plasma Chemistry and Plasma Processing*, vol. 15, no. 4, pp. 711–720, 1995.

[SAM 94] SAMUKAWA S., "Pulse-time-modulated electron cyclotron resonance plasma etching for highly selective, highly anisotropic, and notch-free polycrystalline silicon patterning", *Applied Physics Letters*, vol. 64, no. 25, p. 3398, 1994.

[SHI 12] SHIN H., ZHU W., DONNELLY V.M. *et al.*, "Surprising importance of photo-assisted etching of silicon in chlorine-containing plasmas", *Journal of Vacuum Science & Technology A Vacuum, Surfaces, Film*, vol. 30, no. 2, p. 21306, 2012.

[STE 86] STECKELMACHER W., "Knudsen flow 75 years on: the current state of the art for flow of rarefied gases in tubes and systems", *Reports on Progress in Physics*, vol. 49, p. 1043, 1986.

[STE 89] STEINBRÜCHEL C., "Universal energy dependence of physical and ion-enhanced chemical etch yields at low ion energy", *Applied Physics Letters*, vol. 55, no. 19, pp. 1960–1962, 1989.

[VYV 00] VYVODA M.A., LI M., GRAVES D.B. *et al.*, "Role of sidewall scattering in feature profile evolution during Cl_2 and HBr plasma etching of silicon", *Journal of Vacuum Science & Technology B Microelectronics and Nanometer Structures*, vol. 18, no. 2, p. 820, 2000.

[XU 08] XU L., CHEN L., FUNK M. *et al.*, "Diagnostics of ballistic electrons in a dc/rf hybrid capacitively coupled discharge", *Applied Physics Letters*, vol. 96, no. 26, p. 261502, 2008.

[ZHA 10] ZHANG Q-Z, JIANG W., ZHAO S.-X. *et al.*, "Surface-charging effect of capacitively coupled plasmas driven by combined dc/rf sources", *Journal of Vacuum Science & Technology A Vacuum, Surfaces, Film*, vol. 28, no. 2, p. 287, 2010.

[ZHA 13] ZHAO J.P., CHEN L., FUNK M. *et al.*, "Effect of electron energy distribution functions on plasma generated vacuum ultraviolet in a diffusion plasma excited by a microwave surface wave", *Applied Physics Letters*, vol. 103, no. 3, pp. 2011–2015, 2013.

3

Patterning Challenges in Microelectronics

Patterning is a sequence of process steps that enable the formation of features from a lay-out into a specific material or stack of materials. Typically, patterning is the addition of lithography, plasma etching and stripping or wet cleaning. For decades, optical lithography has been the driver of microelectronics, as the technologies' critical dimensions (CDs) were directly defined by the photolithography resolution known as the Rayleigh criteria. But for the last few years, optical lithography alone has not been able to achieve the higher resolution and lower dimension required for integrated circuits. New patterning techniques such as multiple patterning, EUV lithography, direct self-assembly (DSA), e-beam lithography or nanoimprint lithography are being developed to achieve future node requirements. In this chapter, we will present each alternative technique in terms of the principle, challenges and results from an etch point of view.

3.1. Optical immersion lithography

3.1.1. *Principle*

Photolithography uses a light source projected through a patterned reticle onto a substrate. The basic steps of the photolithographic process are schematically illustrated in Figure 3.1.

Chapter written by Sébastien BARNOLA, Nicolas POSSEME, Stefan LANDIS and Maxime DARNON.

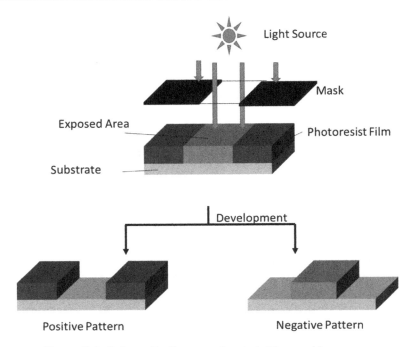

Figure 3.1. *Schematic diagram of a photolithographic process to transfer a positive or negative image onto a substrate*

First, a substrate is coated with a thin film of photosensitive material called a photoresist.

Second, a pattern of light is projected onto the resist film by shining light through a mask on which the required pattern has previously been fabricated, leaving the mask transparent in some areas and opaque in others. The solubility of a photoresist in a particular developer solvent is altered by exposure to radiation, typically light in the ultraviolet (UV) region for photolithography. Therefore, an exposed/unexposed area of a resist film can be removed, leaving a positive/negative tone resist pattern on the substrate after the development process. The resist pattern acts as a temporary mask in the subsequent fabrication process and can be removed after being used.

Optical lithography has a major involvement in the pursuit of increasing transistor density on the wafers, as it is the pattern-defining step. All of the following steps depend on it. Increasing transistor density on a chip requires diminishing the period of the patterns, therefore decreasing the resolution limit.

The resolution capability of photolithography is given by Rayleigh's equation:

$$R = \frac{k_1}{NA} \times \lambda \qquad [3.1]$$

where R is the half-pitch resolution of the image, k_1 is a constant that depends on the resist process and exposure method, λ is the exposure wavelength and NA is the numerical aperture of the projection optic. The numerical aperture of a lens is defined as:

$$NA = n \times \sin\theta \qquad [3.2]$$

where θ is the half-angle of the maximum light cone that can enter or exit the lens, and n is the refractive index of the surrounding medium.

Another important quantity to describe the capability of the optical system is the depth of focus (DOF). This is the range over which an optical image is clear and considered to be in focus. The DOF limits the thickness of the resist film to be exposed and the tolerance in positioning the sample to maintain an image in focus. The general form of the DOF is also derived from the Rayleigh criterion and is expressed as:

$$DOF = k_2 \times \frac{\lambda}{NA^2} \qquad [3.3]$$

where k_2 is a constant found by experiment.

It is clearly seen from equations [3.1] and [3.3] that the resolution and focus capabilities of an exposure tool depend only on wavelength, numerical aperture and constants. These parameters guide how to approach a higher resolution lithography system. The traditional approach is to shorten the radiation wavelength, which is the trend that has been seen for the last decades.

The mercury g-line at 436 nm was the first light source used in photolithography for micrometer-scale resolution. It was succeeded by the mercury i-line at 365 nm and krypton fluoride (KrF) excimer laser at 248 nm. Currently, the most advanced systems operate at 193 nm with an ArF source light.

However, in order to reach a high feature resolution with 193 nm light, it is necessary to significantly improve the NA and reduce the k factor. The numerical aperture of the exposure system has increased from 0.28 to 0.93 with dry and 1.35 with water immersion.

Immersion lithography essentially consists of filling the void between the projecting optic of the scanner and the wafer with a fluid that has a refraction index greater than air (Figure 3.2), which increases the numerical aperture and thus improves the resolution limit. Of course, many issues (e.g. defectivity, tool robustness) due to the use of a fluid (water) in contact with the wafers have been solved to make immersion lithography a viable solution in manufacturing.

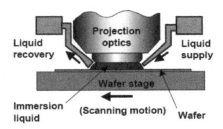

Figure 3.2. *Example of a local wafer immersion system (source: Nikon)*

3.1.2. *Immersion lithography challenges for plasma etching*

Advanced immersion technology extends 193 nm optical lithography to feature sizes below 45 nm by increasing the NA of the photoresist exposure system, but it comes with new challenges. The shrink to the 45 nm node and below has proved especially demanding. Many new processes and materials are being introduced to support the transition. One fundamental challenge for such small features is that DOF and aspect ratio limitations force the resist thickness down to unprecedented levels. For instance, for 45 nm node processes, the resist is less than 150 nm thick and continuously shrinking with new generations. Such a thin imaging layer does not provide sufficient plasma resistance to permit pattern transfer of the image directly to the substrate. The magnification of etch resistance is often required for substrate patterning and is achieved with the use of hard mask technology.

3.1.2.1. Resistance improvement to the etch process

Dry plasma etching is one of the most common pattern transfer techniques used in the IC industry. A patterned resist layer acts as a mask for an underlying substrate such as silicon during the etching process. Although the etch rate of a resist depends on etching gases and process conditions, a resist should have a good resistance to the plasma etching in comparison to that of a substrate. Specifically, the resist should be etched at a slower rate than the substrate material in order to produce an etched structure with a sufficient aspect ratio. Switching from a 248 to 193 nm photoresist can lead to new challenges in terms of patterning. Indeed, the 193 nm photoresist is much more sensitive to the plasma processes. Table 3.1 compares the etch rate of 193 and 248 nm photoresists using typical etch processes [BAZ 07]. Whatever the plasma used, a larger etch rate is observed in the 193 nm than in the 248 nm resist.

Gas	Etch rate (nm/min)	
	248nm PR	193nm PR
O_2	390	419
CF_4	66	122
Cl_2	108	140
HBr	36	68
Ar	24	65

Table 3.1. Photoresist (248 nm vs 193 nm) etch rate comparison as a function of typical etch processes [BAZ 07]

An empirical evolution of the etch rate (ER) with the carbon and oxygen concentration in the photoresist was observed by Ohnishi [GOK 83]. The so-called Ohnishi parameter is defined by:

$$ER \sim \frac{N}{N_c - N_0} \qquad [3.4]$$

where N is the total number of atoms, N_c is the number of carbon atoms and N_0 is the number of oxygen atoms in the polymer.

The Ohnishi parameter is based on the observation that materials containing a high concentration of C–O and C=O groups exhibit much higher sputter yields than pure carbon materials. The "effective carbon content" is defined as the number of carbon atoms minus the number of oxygen atoms normalized to the total number of atoms in the monomer. When the carbon content increases and/or the oxygen content decreases, the Ohnishi parameter decreases and the etch resistance is higher.

Therefore, curing processes have been introduced to improve photoresist behavior before patterning. There are several kinds of plasma processes that have been proposed. Among them, HBr plasmas are the most commonly used. Indeed, it has been demonstrated that vacuum UV (VUV) light from the plasma induces the suppression of lactone and ester groups from the photoresist. This increases the carbon content in the polymer, decreasing the Ohnishi parameter and increasing the etch resistance. This cure also presents the benefit of increasing the mobility of the polymer chains (decrease in glass transition temperature), which leads to surface smoothening to reduce the surface energy of the patterns. The UV-induced resist modification releases carbon-containing groups into the plasma that are eventually dissociated and redeposited on the surfaces exposed to the plasma [BRI 13]. This leads especially to the coating of a graphitized layer on the photoresist surface that prevents lateral etching of the photoresist during the subsequent plasma etching processes, helping in controlling the critical dimension (CD) of the patterns. However, despite this resistance improvement to the etch plasma, this process is not sufficient for the advanced technology nodes that require resist thickness to be thinner. To continue the scaling down, the use of the multilayer hard mask strategy is required.

3.1.2.2. *Multilayer hard mask strategy*

With the thinning of resist thickness as a function of technological node, the use of more than one layer to execute pattern transfer is mandatory. This approach is called the multilayer hard mask strategy. Several different types of processing schemes have been proposed to meet the increased etch resistance demand.

A common multilayer approach is to use alternating organic and inorganic layers, as described in Figure 3.3. In this scheme, the multilayer stack is built on the substrate first by the application of an organic coating, followed by an inorganic coating (typically a silicon-containing material) and then by a resist. This inorganic layer also acts as an anti-reflective layer (also called Si-ARC, Si anti-reflective coating). The organic layer effectively acts as a pattern transfer layer and can also be used to planarize the substrate. It is generally called the OPL (organic planarizing layer), amorphous carbon, SOC (spin on carbon). The organic and inorganic layers can be deposited by PECVD or spin coating.

After the thin resist is patterned with advanced lithographic techniques (Figure 3.3, step 1), the pattern is then transferred into the underlying inorganic layer with a highly selective etch process, typically a fluorinated plasma etch process (Figure 3.3, step 2). The organic layer under the inorganic one is then opened using an oxygen- or hydrogen-based plasma etch process, which takes advantage of the large selectivity achievable between inorganic and organic materials in an oxygen or hydrogen plasma (Figure 3.3, step 3). The substrate is then patterned with the relief image now present in the organic layer (Figure 3.3, step 4).

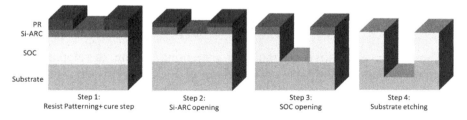

PR			
Si-ARC			
SOC			
Substrate			
Step 1: Resist Patterning+ cure step	Step 2: Si-ARC opening	Step 3: SOC opening	Step 4: Substrate etching

Figure 3.3. *Multilayer hard mask scheme for pattern transfer. Here the inorganic layer is called Si-ARC and the organic one is called SOC*

3.2. Next-generation lithography

Immersion lithography, which has been in reference for more than 10 years, is now stuck to the 80 nm pitch and used in production for half-pitch (hp) 40 nm technologies. The minimum feature size achievable by photolithography is already smaller than the radiation wavelength, and the cost of improving its resolution is increasing exponentially, meaning that it is rapidly becoming economically unviable [HUG 08]. However, the demand for critical feature size reduction continues, and current technology cannot support it. In order to continue miniaturization, a number of alternative technologies are currently being developed. Indeed, as overviewed by the ITRS in 2013, several alternative options are already in production while others are still in the industrialization stage.

To increase the resolution while continuing the use of immersion lithography installed tools, the logical solution was to use double patterning (DP) techniques that divide the minimum feature pitch into two parts. There are several flavors of DP. The lithography-etch, lithography-etch (LELE) approach consists of exposing the wafer twice with a control overlay between the two expositions. The LELE approach can extend the use of immersion lithography in production to the 32 nm technological node. Another flavor of DP called SADP (self-aligned DP) using spacers makes it possible to reach the 20 nm technological node and can be extended to the 16 nm node by repeating the process twice (quadruple patterning). Up until now, there is no clear solution for production below the 16 nm technological node. As multiple patterning is being pushed forward, most of the alternatives   will be competing with this solution that is production proven for less aggressive pitch. The choice for the next technology's patterning solution will be made on performance and cost (Figure 3.4).

The major alternative to multiple patterning is the development of extreme UV lithography with a wavelength of 13.5 nm. However, reducing the wavelength down to such small values necessitates many system changes including vacuum chamber, reflective optics and new light sources that delayed this alternative to sub-10 nm technological node.

Besides DP and EUV lithography, potential next-generation lithography solutions include DSA polymers, nanoimprint and electron-beam (e-beam) lithography.

In the next part, we will discuss the different solutions available in the industry to push the limits of the narrow line patterning and their compatibility with the etching.

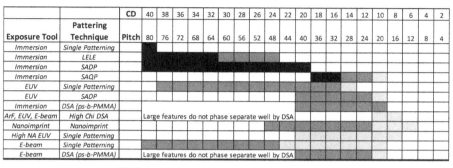

Exposure Tool	Pattering Technique	CD / Pitch																			
		40 / 80	38 / 76	36 / 72	34 / 68	32 / 64	30 / 60	28 / 56	26 / 52	24 / 48	22 / 44	20 / 40	18 / 36	16 / 32	14 / 28	12 / 24	10 / 20	8 / 16	6 / 12	4 / 8	2 / 4
Immersion	Single Patterning																				
Immersion	LELE																				
Immersion	SADP																				
Immersion	SAQP																				
EUV	Single Patterning																				
EUV	SADP																				
Immersion	DSA (ps-b-PMMA)																				
ArF, EUV, E-beam	High Chi DSA	Large features do not phase separate well by DSA																			
Nanoimprint	Nanoimprint																				
High NA EUV	Single Patterning																				
E-beam	Single Patterning																				
E-beam	DSA (ps-b-PMMA)	Large features do not phase separate well by DSA																			

Consense that technique has been used in production
Published demonstrations from potential deployable equipment swho opportunity for production
Simulations, surface images, or research grade demonstration suggest potential for extendability

Figure 3.4. *Potential solution for patterning as a function of CD and pitch. From Neisser et al. [NEI 15]*

3.2.1. *Multiple patterning techniques*

Today, DP is the only manufacturable solution to reach below 40 nm half-pitch, which corresponds to a k_1 factor of 0.3. The extension of this technique to multiple patterning (pitch tripling, pitch quadrupling or pitch octupling) is largely studied today to push immersion lithography to below 10 nm half-pitch [OYA 14].

3.2.1.1. *DP techniques*

DP lithography is one of the simplest emerging next-generation lithographic technologies to implement because it is based on lithographic technology that already exists.

The DP technique splits the required pattern into two steps, in such a way as to separate adjacent features onto separate masks. A sparse feature resolution that is generally higher than a dense feature resolution allows for higher resolution. A 32 nm half-pitch resolution pattern has been demonstrated using DP and 193 nm dry lithography [DAI 08]. An even higher resolution of 22 nm half-pitch has been patterned using a combination of DP and 193 nm immersion lithography [BEN 08]. A variety of DP techniques have been proposed. The most common is called the LELE method and uses two steps of lithography and etching, respectively [DAI 08, BAI 07, DRA 07], as shown schematically in Figure 3.5(a). The first litho-etch step transfers half of the pattern onto a hard mask layer.

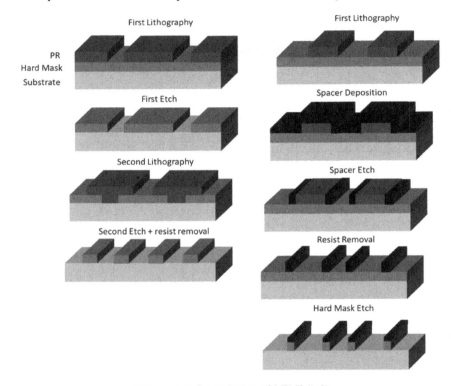

Figure 3.5. *Description of LELE (left) and SADP (right) processes*

The second litho-etch step transfers the other half of the pattern onto the hard mask and the whole pattern is then transferred to the substrate through an etching process. A second DP technique is called self-aligned DP (SADP) [BEN 08, SHI 09] and uses a lithographic pattern itself to position a higher density pattern without the need for advance mask overlaying. The basic SADP process is illustrated in Figure 3.4(b). The resist layer is patterned first (so-called mandrel pattern), followed by deposition of the spacer material. The coating layer is then etched to remove the spacer material on horizontal surfaces and the resist is stripped, leaving just the sidewall spacer forming a masking pattern with twice the pattern density of the lithography-defined original pattern. The spacer pattern is finally transferred onto a hard mask layer underneath. SADP can be extended to give multiple patterning by repeating the deposition of the spacer material and etching steps [CAR 08].

3.2.1.2. Challenges related to multiple patterning techniques

The resolution achievable using the DP technique is promising for the next technology node, but the implementation of DP also doubles the difficulties and manufacturing costs associated with lithographic processes with stringent etch challenges.

3.2.1.2.1. Cost of ownership reduction

Repeating the SADP technique once or twice can be a good solution for extending the immersion 193 nm lithography below 40 nm half-pitch. Self-aligned quadruple patterning and self-aligned octuple patterning integration flows have already been demonstrated for 10 and 5 nm half-pitch, respectively (Figure 3.6). Many efforts are being made to reduce the cost of ownership (COO) of such integration flows. The main ways of improving the cost are: to reduce the number of process steps by innovating the materials that are used for spacer, mandrels and etch stop layers, and by improving the processes themselves. Of course, the etching process improvement is part of this cost reduction strategy [MOH 15].

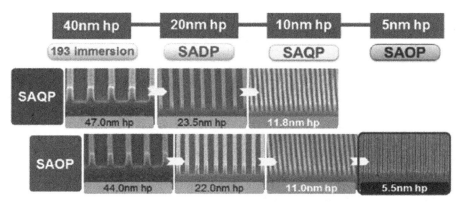

Figure 3.6. *Extension of immersion lithography with SADP, SAQP and SAOP. From Oyama et al. [OYA 14]*

All the multiple patterning techniques need specific litho-etch cut processes to accommodate for the design [OWA 14]. Figure 3.7 shows the evaluated cost comparison for the 14, 10, 7 and 5 nm technology nodes with SADP and SAQP grating lines and litho-etch cutting. At 7 and 5 nm nodes, cost rises rapidly with the increased number of litho-etch cutting. The cost reduction factor is ×0.71 from the 14 nm node to 10 nm node, which can be considered reasonable, while the reduction factors from 10 to 7 nm, and from 7 to 5 nm are higher than ×0.8 (Figure 3.7), which is not cost-effective enough for the semiconductor industry to extend multiple patterning in production for 7 or 5 nm nodes.

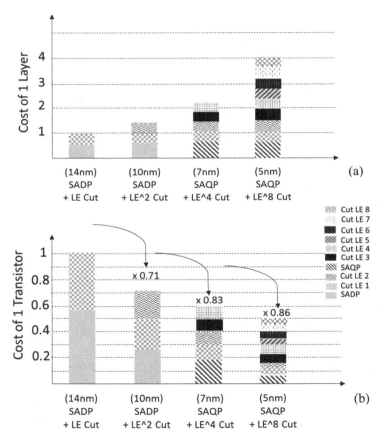

Figure 3.7. *Estimation of cost per layer (a) and transistor (b) with i193 and multiple patterning, down to 5 nm node. From Owa et al. [OWA 14]*

3.2.1.2.2. Multiple patterning etch challenges [BAR 14]

Beyond the economical point of view, multiple patterning strategies also lead to new etch challenges that need to be overcome to achieve the final device. This includes defectivity, CD dispersion, profile dispersion, and line width roughness.

Pattern collapse issue

During the patterning processes, the wafers undergo various cleaning steps that are performed in wet benches. During the drying of the wet chemical, capillary forces on each side of the pattern may be uneven if the patterns are not symmetrical, which leads to lateral forces that bend the pattern. The resulting sway scales with the cube of the pattern aspect ratio [TAN 93]. Therefore, for high aspect ratios, the difference in capillarity forces on each side of the patterns can lead to bending that exceeds the pattern plastic deformation and that can stick to adjacent patterns together. This effect is called pattern collapse and is especially important for high aspect ratio patterns and integration schemes that use wet processes.

Typical SADP processes (see Figure 3.5(b)) use a photoresist or SOC mandrel and SiN (or SiO_2) spacers. This integration allows an easy removal of the mandrel using O_2 plasma, but the Si-ARC needs to be removed before spacer deposition when SOC is used as a mandrel. The fabrication of sub-20 nm wide patterns that are thick enough to enable the transfer by plasma etching leads necessarily to high aspect ratio patterns that are sensitive to pattern collapse.

Figure 3.8(a) reports examples of pattern collapse during wet removal by HF dip of the Si-ARC layer for OPL patterns of 20 nm. From this figure, we can see that adjacent OPL lines are bent and merged. The same process for 25 nm wide patterns (see Figure 3.8(b)) would not lead to pattern collapse since the OPL pattern aspect ratio would be reduced to 5, and therefore the capillary forces would not be able to bend the OPL enough to induce pattern collapse.

Wet cleaning is highly specific to the silicon but cannot be applied to OPL patterns with an aspect ratio larger than 5.

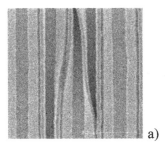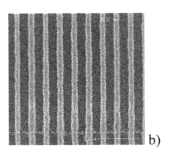

a)   b)

Figure 3.8. *Top CD SEM view after Si-ARC removal using the wet approach (HF 1% 15s) for a 70 nm pitch, a) CD 20 nm (aspect ratio 5) and b) CD 25 nm (aspect ratio 3.8)*

Spacer wiggling

Wiggling is a complex phenomenon induced by feature dimensions (aspect ratio, CD) and the materials' mechanical properties (residual stress, Young's modulus) [DAR 07]. The mechanical residual stress in the films results in line undulation that distorts the patterns. Additional stress from mask chemical modification during pattern transfer may also participate in the wiggling phenomenon [KIM 03b]. For structures with a larger aspect ratio, the mechanical rigidity of the structure may not be large enough to withstand the residual compressive stress, which results in an undulation of the structures. Wiggling for multiple patterning integrations is a direct result of the high aspect ratio of the patterns, as shown in Figure 3.9. Indeed, after silicon etching, important line deformation is observed with an aspect ratio of 7:1 (Figure 3.9(b)), while this effect is not observed for a lower aspect ratio (4:1) (Figure 3.9(c)).

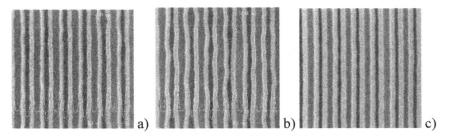

a)   b)   c)

Figure 3.9. *Top CD SEM picture a) before and b) after silicon etching with 10 nm width, 70 nm height (including spacer thickness), and c) with 18 nm width, 70 nm height*

Substrate consumption/microloading issues

During the multiple patterning processes, several etch processes are required, and all the patterns are not necessarily exposed to the same etching processes. For instance, for LELE, the first patterns are etched in the hard mask during a first etching step, and the second patterns are etched during a second etching step. Each etch step may lead to some recess in the substrate and this recess may not be identical. For SADP, the mandrel patterning may also lead to some recess and therefore the space between the spacers (originally filled by the mandrel) has a lower recess than the space defined between two adjacent spacers (originally the space between the mandrels). These differences in recess may lead *in fine* in different pattern profile/depth between the patterns.

In addition, a slight misalignment during LELE would lead to bimodal CD distributions. The same effect would be observed with SADP if the mandrel etch process or the spacer deposition/etching is slightly off target. These nanometer-scale differences would lead to important relative CD variations at narrow dimensions targeted with multiple patterning. For instance, a variation of 1 nm would lead to 10% variations for 10 nm-wide patterns. These variations also result in relatively large aspect ratio variations that eventually lead to pattern profile/depth variations because of aspect ratio-dependent etching. For SADP, the asymmetry of the pattern profile would also lead to odd and even variations of the mask apparent aspect ratio (see Figure 3.10) that would also lead to pattern profile/depth variations.

Figure 3.10 illustrates the odd and even variations of the pattern profile/depth for silicon wire patterning with SADP. The silicon in the space originally filled by the mandrel is hardly etched compared to the silicon between the spacers in the space originally defined by the space between the mandrels. The material of the buffer layer must be carefully chosen to be resistant to either HF dip or fluorocarbon-based plasma and easily removable by wet etch at the end of the process flow.

When a buffer layer is added between the spacers and the silicon layer, this effect is corrected as shown in Figure 3.11.

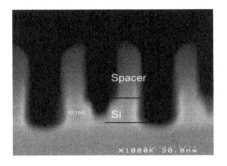 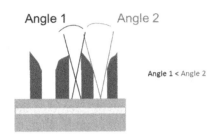

Angle 1 : angle for species collection inside spacers
Angle 2 : angle for species collection outside spacers

Figure 3.10. *Typical issues induced by the integration flow, responsible for CD distortion after Si etching*

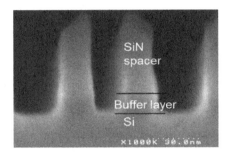

Figure 3.11. *Benefit of introducing a buffer layer between silicon nitride spacer and silicon to limit the microloading issue*

3.2.2. *EUV lithography*

3.2.2.1. *Principle [LAN 11a]*

One of the techniques that replaced 193 nm immersion lithography is extreme ultraviolet lithography (EUVL) using a 13.5 nm wavelength, which has made the most significant progress over the last few years. EUVL was first developed at the end of the 1980s. Since that time, many different studies have been conducted, mainly in the United States, Europe and Japan, in order to identify the feasibility of this new lithographic technique. EUVL is rooted in the continuity of projection optical lithographic techniques currently used in production.

Even if EUV lithography is very similar in its principle to projection optical lithographic technologies, it cannot be considered as a simple extension of the existing technologies, given the specificity of the radiation used and the technological consequences induced by this radiation choice. Indeed, the 13.5 nm wavelength is absorbed by all materials and gases which imposes, on the one hand, the necessity to carry out the whole lithographic process under vacuum and, on the other hand, the need to use reflective optics, including those for mask making. These reflective optical elements are created by stacking many different interferential layers (Bragg mirrors).

3.2.2.2. Challenges to get EUV in production

The introduction of EUVL in a high volume manufacturing (HVM) environment is facing a number of challenges. First, the light source has to deliver a narrow band of high-EUV power (preferably with a high-spectral purity) in order to guarantee high throughput. Second, reticle defectivity has emerged as a major concern for HVM introduction of EUVL. The industry has recognized that a good strategy is essential for maintaining reticle cleanliness as reticle defects induce a major cost penalty on the enablement of EUV.

A key issue in the last 5 years exposure tool light sources, which were not powerful enough to even be used for pilot development. There has been significant progress in the past several years. It is fair to say that EUV scanners now have sufficient productivity to be used for pilot line development of chip-making processes.

The most likely first manufacturing node possible for EUV use is the 7 nm logic node. However, throughput and uptime suitable for manufacturing use still have not been demonstrated. Patterning technology for 7 nm logic node will probably have to be selected sometime in 2016 for production in 2018.

Improvement of scanner throughput, mask defects and scanner uptime are preconditions for successful EUV use in manufacturing. EUV offers substantial benefits in process simplicity compared to multiple patterning by reducing mask counts and allowing more two-dimensional designs to be printed. These benefits will be the driving force for change from multiple patterning. EUV will be adopted when it will be more cost-effective than multiple patterning.

Recent results show that EUV at the 7 nm node can be cost-effective if throughput conditions are met [MAL 14]. A 32% increase in wafer cost is foreseen during a transition from the 16 to 10 nm node based on 193 nm immersion lithography and multiple patterning. This cost is increased by another 14% at the 7 nm technology node. A transition to EUVL at the 7 nm node helps to bring down the wafer cost by 16%, which would bring it on target for a node-on-node cost increase of no more than 20–25%. This, together with scaling benefits, allows the industry to return to Moore's law curve. The sensitivity study of EUV throughput shows that 75 wafers/hour would be the enabler of the technology in terms of cost. EUV will enable a number of critical layers to be printed in single pattern, thus reducing both the number of process steps and the critical masks.

3.2.2.3. Challenges for plasma etching

The adjustment of CD during transfer by reactive ion etch (RIE) of the resist pattern into the subsequent layers, typically an anti-reflective coating (ARC) laid on top of an organic planarizing layer (OPL), is well established for 193 nm lithography. By optimizing the ARC and OPL trimming, it is possible to modulate the CD before the final transfer on the wafer. But with the incoming thickness reduction of EUV resist, most of our current shrink knowledge is ineffective or insufficient.

The major challenge is to improve the selectivity of resist over the transfer layer while minimizing LER (line edge roughness) and LWR (line width roughness). As the LER of EUV resist is about 50% higher than for ArF resist (Figure 3.12(a)), roughness mitigation is a major driver for subsequent plasma etching steps. As for 193 nm resist, different plasma pretreatments are able to smooth the resist before being transferred (Figure 3.12(b)).

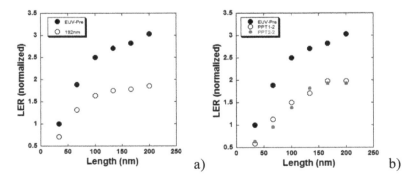

Figure 3.12. a) LER comparison between ArF and EUV PR and b) LER improvement with plasma pretreatment (PPT). From Lee et al. [LEE 15]

The other concern is the selectivity of EUV PR that needs to be improved. Several solutions have been implemented on etch tools such as using pulsed plasma [LEE 15] or DC superposition [HON 12], jointly with specific chemistries. Indeed, material challenges stimulate the introduction of new materials such as inorganic-based resist [KRY 14] or metal-based hard mask materials [GUE 14]. All of these new solutions, if they are implemented, may impact the plasma etching processes in order to enable the use of EUV in production.

3.2.3. E-beam lithography [LAN 11a]

3.2.3.1. Principles and challenges

The first electron beam lithography tools were developed in the late 1960s. They were based on the principle of scanning electron microscopes (SEMs). Since its introduction, electron beam (or e-beam, as it is known) has been widely used in laboratories and universities because of its strong capacities in terms of resolution and flexibility. In the 1970s, IBM developed the shaped beam concept, which brought significant enhancements to the writing speed, an essential step for industry in order to achieve implementation of high-throughput tools. However, this new technology, called single beam technology, is several times slower than mask-based optical lithography. Therefore, logically, its industrial attractiveness rapidly declined because of its production costs, which were higher than those of optical lithography. Since the early 2000s and the introduction of 193 nm, the cost of optical lithography has risen significantly with each new generation, due to the increase in the cost of the 193 nm tool and the intensive use of optical lithography-associated techniques to enhance the resolution (resolution-enhanced techniques). In this context, inserting new e-beam tools for direct writing and high throughput becomes a strong potential alternative to reduce industrial costs related to the lithographic process steps. This renewal of interest is particularly encouraged by application-specific integrated circuit manufacturers and foundries, for which the mask budget represents a large part of total costs. Maskless lithography was then foreseen in the ITRS roadmap for the 32 nm technological node. Indeed, with the development of multibeam tools to increase productivity, e-beam lithography became a serious option in the race to scaling.

3.2.3.2. *The Mapper solution*

In principle, the technology uses the principle of parallel beams from a unique source, driven by a data transport system that carries the information to the deflectors. The industrial platform plans to use 13,000 individually controlled beams, which will cover a width of 26 mm (standard optical scanner field size). Each beam writes a 2 μm-wide band perpendicular to the direction of the table, with a slight overlap with each nearest beam. Circuit writing data are generated on a server system separate to the exposure platform and transmitted optically by 13,000 light beams (Figure 3.13(b)), each of which control the deflection of a beam. The deflectors and switches of the sub-beams are controlled by integrated photosensors. The optical control makes it possible to address the sub-beams at a very high frequency of a few gigahertz. Mapper lithography's published roadmap shows the development of a unit with a 10 wafer-per-hour throughput using 13,260 beams targeted at various layers for the 14, 10 and 7 nm logic nodes.

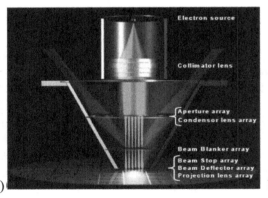

Figure 3.13. *a) Photograph of pre-alpha multibeam tool installed at CEA-LETI and b) schematic of the Mapper solution. From Servin et al. [SER 15]*

E-beam Direct Write (EBDW) is still on the way to be manufacturing-friendly, but its extendibility down to a half-pitch of 16 nm has already been demonstrated (Figure 3.14).

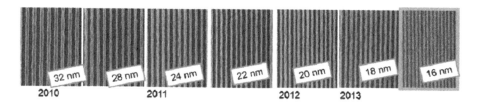

Figure 3.14. *EBDW scalability demonstrated down to a half-pitch of 16 nm. From Servin et al. [SER 15]*

Its compatibility with CAR resist and materials such as trilayer, enabling the use of regular plasma etching strategy, is shown in Figure 3.15.

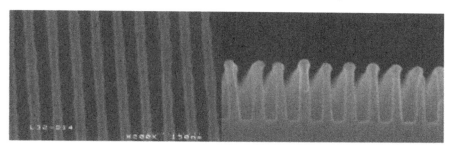

Figure 3.15. *Transfer of 32 nm half-pitch L/S features into trilayer stack. From Servin et al. [SER 15]*

3.2.4. *Direct self-assembly*

3.2.4.1. *Principle of DSA*

3.2.4.1.1. Block copolymer materials (BCP) [LAN 11b]

The use of block copolymer (BCP) thin films seems to be a powerful alternative for specific applications to overcome the intrinsic limitations of traditional lithographic techniques. Based on the self-organization of polymeric chains similar to conventional polymer photoresist chains used in semiconductor fabrication, this technology allows the realization of regular patterns whose dimensions cannot be achieved by optical lithographic processes. After decades of being only secondary in importance, as reflected by an almost negligible number of publications on the subject, the capacity of BCPs to self-assemble into periodic morphologies is likely to play a key role in nanotechnological applications in the future. The use of the BCP technique to obtain lithographic masks was first proposed in 1995 by Mansky [MAN 95], and appears today as one of the technological solutions to generate localized uniform objects with low dimension (~20 nm) and high density (~10^{11}/cm^2). Block copolymers are a specific type of polymer and can be classified as soft matter. They are composed of at least two chemically different polymer fragments ("blocks") that are covalently linked. Generally, copolymers are flexible and can bend at monomer junctions, leading to a great architectural diversity, examples of which include linear copolymers, graft copolymers and star copolymers. When only two sub-chains of different monomers, A and B, are bound covalently to each other, a diblock copolymer is formed. A simple linear AB diblock copolymer chain (called: PA-b-PB) is composed of fN polymer A segments and $(1-f)N$ polymer B segments, linked together at one end by a covalent bond, where N denotes the degree of polymerization and f the fraction of block A in the chain. Depending on the composition, f, diblock copolymers can organize into several equilibrium structures: lamellae, hexagonally ordered nanodomains, bicontinuous cubic gyroids or a body-centered cubic lattice of spheres (see Figure 3.16).

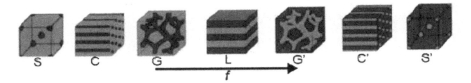

Figure 3.16. *Schematic representation of diblock copolymer phases as a function of f: (S) spherical; (C) cylindrical; (G) gyroid; (L) lamellar; (G') inverse-gyroid; (C') inverse-cylindrical; (S') inverse-spherical phases*

3.2.4.1.2. PS-b-PMMA polymers

Among the various BCP systems reported in the literature, poly(styrene)-block-poly(methyl methacrylate) (denoted as PS-b-PMMA) is well studied, and presents interesting properties from a technological point of view. Firstly, styrene and methyl methacrylate monomers can be readily copolymerized without a metal catalyst in order to obtain a random copolymer (denoted as PS-r-PMMA) with precise composition. This property is extremely valuable since it is limited to only a few monomeric systems, and because random copolymers are used as an effective "neutral" passivation layer on substrates to properly orient the BCP features. Moreover, the intrinsic chemical nature of both PS and PMMA monomers allows, under appropriate conditions, to synthesize a BCP with well-controlled architecture and composition, and low disparity of chain distribution. Consequently, perpendicular BCP features can be obtained without the use of sophisticated systems (solvents atmosphere, vacuum, etc.), which could be potentially prohibitive for lithography. It has also been demonstrated that the self-assembly of a PS-b-PMMA system takes place with a simple thermal bake at an elevated temperature in a short time scale, i.e. under similar conditions as classical lithographic processes. Low-molecular-weight PS-b-PMMA systems can reach perpendicular features with CDs as low as ~10 nm (Figure 3.17), therefore making such materials very attractive to the corresponding technological node. Conversely, high-molecular-weight BCPs can also be synthesized efficiently to achieve features with natural periods close to ~50 nm, rendering this kind of polymer interesting for an early introduction of BCP technology into the current logical node step [CHE 13].

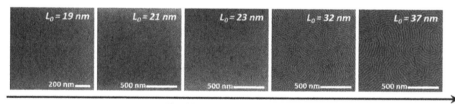

Lamellar block-copolymer period

Figure 3.17. *SEM pictures showing the perpendicular self-assembly of lamellar BCPs with periods ranging from 19 up to 37 nm. From Chevalier et al. [CHE 13]*

3.2.4.1.3. High χi materials

Despite the fact that PS-b-PMMA polymers present interesting properties, achievable CDs with this system will not decrease much below 10 nm due to its relatively low value of the Flory–Huggins parameter (χ). To address this specific problem, BCP chemical systems with higher χ values have to be developed, along with their self-assembly processes. The SEM image in Figure 3.18 shows that BCP features close to 7 nm may be reached with such new polymer systems, and that they can be efficiently guided with the traditional grapho-epitaxy approach. These results indicate the usefulness and capabilities of BCP materials to scale down dimensions of future interest for lithographic applications.

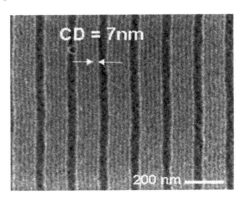

Figure 3.18. *SEM pictures of high-χ BCPs self-assembled within the grapho-epitaxy approach (HSQ templates) leading to features down to 7 nm. From Chevalier et al. [CHE 13]*

3.2.4.1.4. Integration of BCP materials

Various approaches have been studied over several years in order to improve the correlation length of the BCP self-assembly. Among them, the so-called chemi-epitaxy, which uses patterns based on chemical contrasts [KIM 03a], and the grapho-epitaxy [SEG 01], which uses physical templates (e.g. lithographic resists hardwalls), are both well-known methods to efficiently guide the BCP self-assembly. The improvement of these two techniques leads to their recent introduction in 300 mm tracks as BCP test structures, chemi-epitaxy (Figure 3.19), and grapho-epitaxy for line–space applications in lithography (Figure 3.20).

One chemo-epitaxy integration scheme is based on a "lift-off" process flow, as shown in Figure 3.19 [RAT 12]. In the lift-off process, immersion lithography is used to pattern a positive tone resist on a spin-coated silicon BARC. The polarity of the developed lines is switched using a UV exposure and a bake step, which makes the pre-pattern soluble in the developer. A neutral layer is then used to coat the top of the polarity switched pre-pattern. The TMAH developer is used to diffuse through the neutral layer material and lift off or solubilize the pre-pattern resist. This generates a chemical substrate containing silicon BARC guide structures interlaced with neutral layer stripes. A PS-b-PMMA BCP is coated over the substrate and annealed to generate vertical lamella domains that can be etched into lithographic patterns.

After optimizing the process, 14 nm half-pitch PS-b-PMMA patterns were demonstrated using a 4× multiplication process from an immersion pre-pattern (35 nm lines/112 nm pitch), as shown in Figure 3.19 [RAT 12].

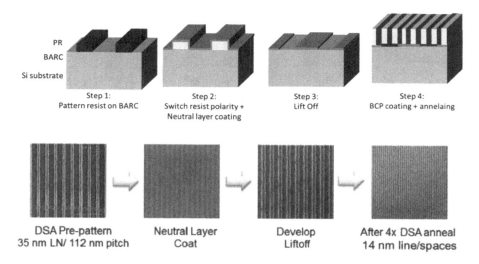

Figure 3.19. *Example of chemo-epitaxy integration flow for line and space features. From Rathsack et al. [RAT 12]*

Grapho-epitaxy uses physical pre-pattern guides or templates to array DSA patterns. One of the integration schemes is to utilize a guide structure formed with a negative tone-developed pre-pattern followed by trilayer opening, as shown in Figure 3.20.

First, steps 1 to 2 show how using a dry exposure at 193 nm from scanner equipment patterns are generated into an NTD positive and transferred by means of dry plasma etching into an inorganic (Si-ARC) and organic (SOC) hard masks. During step 3, spin-coating and grafting of "brush" materials imprints its affinity to the template with a thin (few nm) polymer layer. In all configurations, the DSA process is next in step 4: the BCP is spin-coated at a controlled thickness that fills patterns to the top; after annealing at a temperature >200 °C for a few minutes, the self-assembly is complete and lamellae appear in the cavity (step 5).

After optimizing the process, PS-b-PMMA patterns are demonstrated using a 4× multiplication process from a trilayer pattern (90 nm lines/180 nm pitch), as shown in Figure 3.20 [CLA 16].

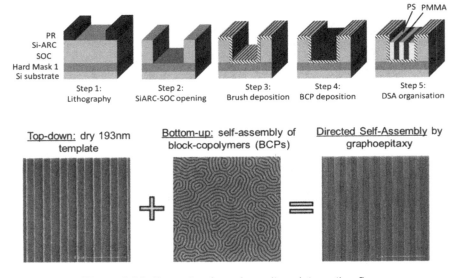

Figure 3.20. *Example of grapho-epitaxy integration flow for line and space features. From Claveau et al. [CLA 16]*

3.2.4.1.5. DSA challenges for plasma etching

One of the major challenges for DSA integration concerns the removal of PMMA selectively to PS. Indeed, it is important to remove the PMMA without altering the PS in order to be able to transfer patterns from the BCP into subjacent layers. This step is very challenging since PS and PMMA present a similar bone architecture, and thus achieving a high selectivity between the two is very challenging. Today, two approaches are investigated for the removal of PMMA: wet or plasma etching.

The first approach consists of using a solvent such as acetic acid. It has been demonstrated that acetic acid provides an infinite selectivity between these polymers so that PMMA can be removed without consuming PS [SUZ 12]. However, line pattern collapse, due to capillarity forces induced during the solvent development, has been observed on lamellar BCP with an aspect ratio over 1 [PIM 14].

Plasma provides better compatibility with narrow lines, limiting the risk of pattern collapse for lamellar applications. However, one major drawback of this technique is the difficulty in achieving high selectivity between PMMA and PS films. Selectivity between PMMA and PS can be expected because the C/O ratio concentration is higher for PS than for PMMA [GOK 83] and the carboxylic compound is not as resistant to the plasma etching as the phenyl group [OEH 11].

Previously, O_2-based chemistries were developed to remove PMMA [LIU 10]. But their CD control was very challenging since the etching was very reactive [FAR 10, DEL 14], and presented low selectivity (about 2). Therefore, new chemistries have been proposed in the literature. For example, selectivity has been improved by adding a PS protection step: Chan *et al.* proposed the addition of an argon step after the Ar-O_2 one [CHA 13]. It has been shown that the addition of pure argon provides a higher PMMA:PS selectivity (about 16:1 on blanket wafers). On lamellar PS-b-PMMA films, this argon step is used for protecting PS with a non-volatile product from PMMA that are deposited on the top of PS patterns during PMMA removal [CHA 14]. For removing the whole PMMA pattern, Ar-O_2 chemistry is used. This improvement provides a better PMMA:PS selectivity than Ar-O_2 chemistry and a lower roughness on lamellar copolymers. Nevertheless, the selectivity is lower on PS-b-PMMA films than on blanket wafers (only 4:1 on patterns).

Alternative chemistries (CO, Xe, H_2) have shown a higher selectivity potential. Omura *et al.* proposed the use of CO chemistry because it provides an infinite selectivity during the first few seconds [OMU 14]. A saturation phenomenon is then observed for PMMA consumption. It has been shown that CO provides infinite selectivity to PS, but an etch stop phenomenon is observed on PMMA after a few seconds of etching [SAR 16]. Xe or H_2 addition to CO chemistry fully etches the PMMA but with lower selectivity to PS (see Figure 3.21) [SAR 16].

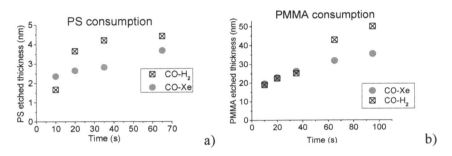

Figure 3.21. *a) PS and b) PMMA consumption evolution on blanket wafer with the process time for CO-H$_2$ (1:1) and CO-Xe (1:1). From Sarrazin et al. [SAR 16]*

Thus, a combination of CO with Xe or H_2 has been investigated for a trade-off between process speed and selectivity. The benefit of CO-H$_2$ chemistry was demonstrated by Sarrazin *et al.* [SAR 16], showing a good PMMA:PS selectivity (12:1), a high PMMA etch rate and a low PS degradation (Figure 3.22).

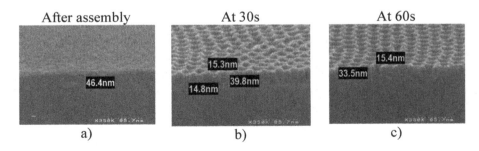

Figure 3.22. *Cross-sectional SEM images of cylindrical patterns after a) self-assembly and b) CO-H2 30s and c) 60s process times. From Sarrazin et al. [SAR 16]*

3.2.5. *Nanoimprint lithography (NIL)*

3.2.5.1. *Principle and challenges [LAN 11b]*

Nanoimprint is a generic technology involving various approaches but with a common goal: the use of a stamp or mold to transfer a 2D or 3D pattern onto a surface or in a material's thickness. All of these technologies assume a close contact between the original information media (the mold) and the receiving support (the substrate). Two main NIL technologies had the biggest impact on micro- and nanofabrication of electronic, optical, biological or magnetic devices: thermally and UV-assisted nanoimprint lithography (Figure 3.23). Nanoimprint techniques stand out from the other more conventional lithographic processes (photolithography, electronic lithography, X-ray, EUV lithography, etc.) because of the fundamental mechanism of creating the structures. In classical approaches, structures are created through a physical and chemical contrast. The resist, if it is positive, can be selectively developed and inversely for the negative resist. In the case of nanoimprint, the contrast is topographic and the flow of the resist through the stamp's cavities shapes the pattern. Based on this principle, two solutions are used to solidify the displaced matter. Either the thermal properties of the material are used (in thermally assisted nanoimprint) by exploiting its liquid and solid states, or its physical and chemical properties are thermally modified (with thermoset materials) or modified by exposure to utlraviolet (UV-assisted nanoimprint), establishing a cross-linkage or polymerization process.

The key element of the process, the mold or stamp, is to nanoimprint what the mask is to optical lithography. However, unlike projection optical lithography, nanoimprint is a 1X technology because there is no reduction factor between the mask and the resist. Therefore, the structures reproduced in the resist have the same size as those on the mold. The conception of such an object requires advanced micro- and nanofabrication processes (mainly etching and lithography). The most common approach involves the use of other highly resolving lithographic techniques for stamp manufacturing such as electron beam lithography, EUV, X-rays, advanced optical (193 nm dry immersion lithography) or even FIB lithography.

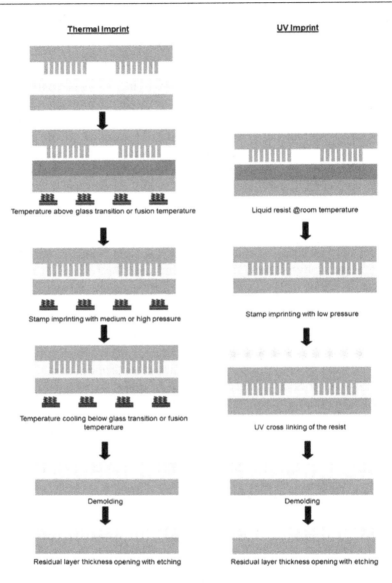

Figure 3.23. *Description of thermally and UV-assisted nanoimprint processes. From Landis et al. [LAN 11b]*

3.2.5.2. *Applications of nanoimprint lithography*

Nanoimprint lithography has been recently evaluated as a cost-effective alternative to multiple patterning and EUV for memory applications [TAK 15]. Most of the challenges (overlay, CDU, defectivity, throughput) were improved to make NIL a real alternative for non-volatile memory manufacturing. Overlay was improved by a factor of 8 in 2 years, and sub-5 nm 3 σ overlay was demonstrated. Throughput per imprint station was improved fivefold, i.e. to 10 wafers per hour. Overall process defectivity was reduced by two orders of magnitude, i.e. to 9 defects per cm². Defectivity is not yet at the desired level for manufacturing flash memory, but progress is being made at such a rate that it will reach the target level by 2016 or 2017. Cost is the biggest driver for NIL as it can induce a reduction of 40% compared to SAQP if used in production for 15 nm half-pitch flash memory technology, as shown in Figure 3.24.

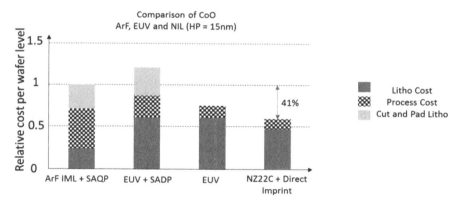

Figure 3.24. *Comparison of cost per wafer level for 15 nm half-pitch memory technology [TAK 15]*

NIL also has a lot of values for "More than Moore" applications such as LED, bio-MEMS and display, either for low-cost aspects or for specific lithography capabilities, such as the unique capability to print 3D resist profiles (Figure 3.25). In all these applications, a very close coupling is required between NIL and etching to achieve the final feature profiles.

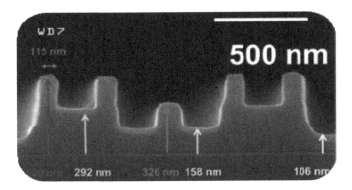

Figure 3.25. *Cross-sectional SEM image of typical 3D resist profiles after NIL printing (source: CEA-LETI)*

3.3. Conclusion

The conventional UV lithographic technologies used for years in microelectronics have reached their limit and new patterning technologies are now in production and/or under consideration for future technological nodes. This includes the development of multiple patterning technologies such as LELE, SADP or self-aligned quadruple patterning. Extreme UV lithography with reflective optics is also under consideration for the sub-10 nm technological nodes. Electron beam lithography, DSA and nanoimprint that have so far been confined to research laboratories are now seriously being considered as challengers for future technological nodes.

With the introduction of alternative lithographic techniques, plasma etching processes are being challenged more than ever, with requirements below the nanometer scale in terms of CD control, etch stop layer consumption, profiles, line edge roughness, aspect ratio and new materials such as III–V, ferro-magnetic materials or polymers. The patterning solution directly conditions the etching process and *vice versa*, making the connection between lithography and plasma etching compulsory to choose the most appropriate patterning technique.

Such aggressive requirements require significant innovation at several levels: plasma sources (low Te, pulsed plasma), chamber hardware (fine center to edge uniformity knobs), chemistries (new gases or combination of gases) and specific process control hardware (precise temperature probes, new plasma sensors). This also increases the complexity of process development to an unprecedented level.

3.4. Bibliography

[BAI 07] BAILEY G.E., TRITCHKOV A., PARK J.-W. *et al.*, "Double pattern EDA solution for 32 nm HP and beyond", *Proceedings of SPIE*, vol. 6521, 65211K, 2007.

[BAR 14] BARNOLA S., PIMENTA BARROS P., ARVET C. *et al.*, "Plasma etching and integration challenges using alternative patterning techniques for 11 nm node and beyond", *Proceedings of SPIE*, vol. 9054, 2014.

[BAZ 07] BAZIN A., MAY M., PARGON E. *et al.*, "Study of 193-nm resist degradation under various etch chemistries", *Proceedings of SPIE*, vol. 65192N, 2007.

[BEN 08] BENCHER C., CHEN Y., DAI H. *et al.*, "22 nm half-pitch patterning by CVD spacer self alignment double patterning (SADP)", *Proceedings of SPIE*, vol. 6924, 65244E, 2008.

[BRI 13] BRIHOUM M., RAMOS R., MENGUELTI L. *et al.*, "Revisiting the mechanisms involved in Line Width Roughness smoothing of 193 nm photoresist patterns during HBr plasma treatment", *Journal of Applied Physics*, vol. 113, p. 013302, 2013.

[CAR 08] CARLSON A., KING LIU T.-J., "Negative and iterated spacer lithography processes for low variability and ultra-dense integration", *Proceedings of SPIE*, vol. 6924, 69240B, 2008.

[CHA 13] CHAN, B., TAHARA S., "Etching method using block-copolymers", Patents US20140131839 A1, 2014.

[CHA 14] CHAN B.T., TAHARA S., PARNELL D. *et al.*, "28nm pitch of line/space pattern transfer into silicon substrates with chemo-epitaxy Directed Self-Assembly (DSA) process flow", *Microelectronic Engineering*, vol. 123, pp. 180–186, 2014.

[CHE 13] CHEVALIER X., NICOLET C., TIRON R. *et al.*, "Scaling-down lithographic dimensions with blockcopolymer materials: 10-nm-sized features with poly (styrene)-block-poly(methylmethacrylate)", *Journal of Micro/Nanolithography MEMS MOEMS*, vol. 12, no. 3, p. 031102, 2013.

[CLA 16] CLAVEAU G., QUEMERE P., ARGOUD M. *et al.*, "Surface affinity role in directed self-assembly of lamellar block copolymers", *Proceedings of SPIE*, vol. 9779, 2016.

[DAI 08] DAI H., BENCHER C., CHEN Y. et al., "45 nm and 32 nm half-pitch with 193 dry lithography and double patterning", Proceedings of SPIE, vol. 6924, 652421, 2008.

[DAR 07] DARNON M., CHEVOLLEAU T., JOUBERT O. et al., "Undulation of sub-100 nm porous dielectric structures: a mechanical analysis", Applied Physics Letters, vol. 91, p. 193103, 2007.

[DEL 14] DELALANDE M., CUNGE G., CHEVOLLEAU T. et al., "Development of plasma etching processes to pattern sub-15nm features with PS-b-PMMA block copolymer masks: application to advanced CMOS technology", Journal of Vacuum Science and Technology B, 32, p. 051806, 2014.

[DRA 07] DRAPEAU M., WIAUS V., HENDRICKX E. et al., "Double patterning design split implementation and validation for the 32 nm node", Proceedings of SPIE, vol. 6521, 652109, 2007.

[FAR 10] FARRELL R.A., PETKOV N., SHAW M. et al., "Monitoring PMMA elimination by reactive ion etching from a Lamellar PS-b-PMMA thin film by ex situ TEM methods", Macromolecules, vol. 43, no. 20, pp. 8651–8655, 2010.

[GOK 83] GOKEN H., ESHO S., OHNISHI Y., "Dry etch resistance of organic materials", J. Electrochem. Soc., 130, pp. 143–146, 1983.

[GUE 14] GUERRERO D. J., "Extending lithography with advanced materials", Proceedings of SPIE, vol. 9051, 2014.

[HAZ 09] HAZELTON A., WAKAMOTO S., HIRUKAWA S. et al., "Double patterning requirements for optical lithography and prospects for optical extension without double patterning", J. Micro/Nanolith. MEMS MOEMS, vol. 8, no. 011003, pp. 1–11, 2009.

[HON 12] HONA M., YATSUDA K., "Patterning enhancement techniques by reactive ion etch", Proceedings of SPIE, vol. 8328, 2012.

[HUG 08] HUGHES G., LITT L.C., WÜEST A. et al., "Mask and wafer cost of ownership (COO) from 65 to 22 nm half-pitch nodes", Proceedings of SPIE, vol. 7028, 70281P, 2008.

[KIM 03a] KIM S.O, SOLAK H.H, STOYKOVICH M. et al., "Epitaxial self-assembly of block copolymers on lithographically defined nanopatterned substrate", Nature vol. 424, no. 6947, pp. 411–414, 2003.

[KIM 03b] KIM J., CHAE YS, LEE WS, et al., "Sub-0.1 µm nitride hard mask open process without procuring the ArF photoresist", Journal of Vacuum Science and Technology B, vol. 21, p. 790, 2003.

[KRY 14] KRYSAK M.E, BLACKWELL J.M, PUTNA S.E, *et al.*, "Investigation of novel inorganic resists materials for EUV lithography", *Proceedings of the SPIE*, vol. 9048, 2014.

[LAN 11a] LANDIS S., *Lithography*, ISTE Ltd, London and John Wiley & Sons, New York, 2011.

[LAN 11b] LANDIS S., *Nano-lithography*, ISTE Ltd, London and John Wiley & Sons, New York, 2011.

[LEE 15] LEE C., NAGABHIRAVA B., GOSS M. *et al.*, "Plasma etch challenges with new EUV lithography material introduction for patterning for MOL and BEOL", *Proceedings of SPIE*, vol. 9428, 2015.

[LIU 10] LIU C., LIU C., NEALEY P. *et al.*, Integration of block copolymer directed assembly with 193 immersion lithography, *Journal of Vacuum Science and Technology B*, vol. 28, pp. C6B30–C6B34, 2010.

[MAL 14] MALLIK A., HORIGUCHI N., BÖMMELS J. *et al.*, "The economic impact of EUV lithography on critical process modules", *Proceedings of SPIE*, vol. 9048, 2014.

[MAN 95] MANSKY P., CHAIKIN, THOMAS E., "Monolayer films of diblock copolymer microdomains for nanolithographic applications", *Journal of Materials Science*, vol. 30, p. 1987–1992, 1995.

[MOH 15] MOHANTY N., FRANKE E., LIU C. *et al.*, "Challenges and mitigation strategies for resist trim etch in resist mandrel based SAQP integration scheme", *Proceedings of SPIE*, vol. 9428, 2015.

[NEI 15] NEISSER M., WURM S., "ITRS lithography roadmap: 2015 challenges", *Advances in Optical Technologies*, vol. 4, no. 4, pp. 235–240, 2015.

[OEH 11] OEHRLEIN G.S., PHANEUF R., GRAVES D., "Plasma-polymer interactions: a review of progress in understanding polymer resist mask durability during plasma etching for nanoscale fabrication", *Journal of Vacuum Science and Technology B*, vol. 29, p. 010801, 2011.

[OMU 14] OMURA M., IMAMURA T., YAMAMOTO H. *et al.*, "Highly selective etch gas chemistry design for precise DSAL dry development process", *Proceedings of SPIE*, 9054, 9, 2014.

[OWA 14] OWA S., WAKAMOTO S., MURAYAMA M. *et al.*, "Immersion lithography extension to sub-10 nm nodes with multiple patterning", *Proceedings of SPIE*, vol. 9052, 2014.

[OYA 14] OYAMA K., YAMAUCHI S., NATORI S. *et al.*, "Robust complementary technique with multiple-patterning for sub-10 nm node device", *Proceedings of SPIE*, vol. 9051, 2014.

[PIM 14] PIMENTA-BARROS P., BARNOLA S., GHARBI A. *et al.*, "Etch challenges for DSA implementation in CMOS via patterning", *Proceedings of SPIE*, vol. 9054, 15, 2014.

[RAT 12] RATHSACK S., SOMERVELL M., HOOGEA J. *et al.*, "Pattern scaling with directed self assembly through lithography and etch process integration", *Proceedings of SPIE*, vol. 8323, 2012.

[SAR 16] SARRAZIN A., POSSEME N., PIMENTA-BARROS P. *et al.*, "PMMA removal selectivity to PS using dry etch approach", *J. Vac. Sci. Technol. A*, vol. 34, no. 6, 2016.

[SEG 01] SEGALMAN R.A., YOKOYAMA H., KRAMER E.J., "Graphoepitaxy of spherical domain block copolymer films", *Advanced Materials*, vol. 13, no. 15, pp. 1152–1155, 2001.

[SER 15] SERVIN I., THIAM N., PIMENTA-BARROS P. *et al.*, "Ready for multi-beam exposure at 5kV on MAPPER tool: lithographic & process integration performances of advanced resists/stack", *Proceedings of the SPIE*, vol. 9423, 2015.

[SHI 09] SHIU W., LIU J.H., WU J.S. *et al.*, "Advanced self-aligned double patterning development for sub-30-nm DRAM manufacturing", *Proceedings of SPIE*, vol. 7274, 72740E, 2009.

[SUZ 12] SUZUKI H., KOMETANI R., ISHIHARA S, *et al.*, "Selectively patterned metal nanodots fabrication by combining block copolymer self-assembly and electron beam lithography", *Proceedings of SPIE*, vol. 8463, 6, 2012.

[TAK 15] TAKEISHI H., SREENIVASAN S.V., "Nanoimprint system development and status for high volume semiconductor manufacturing", *Proceedings of SPIE*, vol. 9423, 2015.

[TAN 93] TANAKA T., MORIGAMI M., ATODA N., "Mechanism of resist pattern collapse during development process", *Japanese Journal of Applied Physics*, vol. 32, pp. 6059–6064, 1993.

4

Plasma Etch Challenges
for Gate Patterning

The patterning of the gate electrode is a critical step for transistor fabrication. For years, doped polysilicon (pSi) has been used as a gate metal while silicon dioxide has been used as a gate dielectric. After SiO_2 and pSi deposition, the gate was patterned by conventional lithography and plasma etching (described in Chapters 3 and 2, respectively). At each technological generation, the length of the gate and the thickness of the gate dielectric were reduced. However, when the thickness of the gate oxide was reduced below a few nanometers, the tunneling current increased. In addition, high-performance CMOS circuits take advantage of different work functions for the gate metals between P and N transistors. For all these reasons, the standard doped pSi/SiO_2 structures shifted towards dual metal (different metal for N and P transistors) high-k dielectric structures. Initially, the dual metal was obtained with different dopings of pSi, and, today, it is obtained by integrating different metal materials between the N and P transistors. The high-k dielectrics were initially based on $SiON/SiO_2$ layers and then shifted to hafnium-based dielectrics (HfO_2, HfSiON, etc.).

Along with this change of structure came an alternative gate patterning technique based on a "replacement-gate" or "gate-last" approach. In this case,

---

Chapter written by Maxime DARNON and Nicolas POSSEME.

a standard dummy pSi/SiO$_2$ gate structure is patterned with conventional techniques. The complete transistor is built with spacers' deposition (forming the mask for source/drain implantation, as well as encapsulating and protecting the sidewalls of the gate), source/drain doping, silicide formation and isolation. Finally, the dummy gate is removed and replaced by the real gate structure. There are several approaches, depending on whether the gate oxide is also replaced or not.

Figure 4.1 illustrates the gate-first and gate-last approaches. The major advantage of the gate-last approach is to avoid the exposure of gate materials to the high-temperature steps used for doping and silicide formation. However, this approach adds other constraints and both approaches (gate-first and gate-last) are currently integrated in high volume manufacturing.

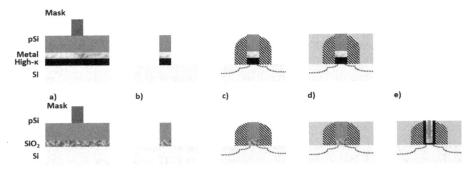

Figure 4.1. *Simplified schematic view of the gate patterning with the gate-first (top) and gate-last (bottom) approaches. a) Material deposition and mask patterning; b) (dummy) gate patterning; c) spacers' formation and doping; d) pre-metal dielectric deposition and planarization; e) dummy gate removal and gate material deposition and planarization*

Regardless of the gate patterning strategy, gate patterning requires the perfect transfer of very narrow structures (typically 20 nm at the 14 nm node according to the International Technology Roadmap for Semiconductor (ITRS)) from a mask material (typically SiO$_2$ or amorphous carbon) to pSi, metals and eventually a high-k dielectric, while avoiding any damage in the source/drain region. These gate etch challenges and associated solutions are discussed below.

4.1. pSi gate etching

Silicon etching is performed with halogen-based plasmas. There are two competing approaches: fluorine-based processes and chlorine/bromine-based processes.

4.1.1. Etching processes in HBr/Cl₂/O₂

Traditional processes for gate patterning are based on chlorine-based plasmas. In these processes, bowed profiles and microtrenching are frequently observed [DES 00]. The addition of HBr in the plasma helps to straighten the profile and prevents the formation of microtrenching [DES 00, LAN 00]. The addition of oxygen improves the selectivity with the gate oxide, and strengthens the passivation layers at the pattern sidewalls [BEL 96, BEL 97, CUN 02]. These processes are better suited for oxide masks than for carbon-based masks because of the high reactivity of oxygen with carbon [BEL 97]. The profile of the gate in such plasmas is controlled by the passivation layers formed at the pattern sidewalls. As shown in Figure 4.2, triangular-shaped passivation layers form at the pattern sidewalls during etching, and there is a direct correlation between the passivation layer thickness and the gate dimension [DET 03]. Passivation layers are formed from the oxidation of silicon-containing species at the pattern sidewalls [CUN 02]. Indeed, volatile silicon-containing etch by-products are formed at the etch front and feed the plasma. These species can stick on all the surfaces, but their high volatility and etching by plasma radicals lead to a short retention time before desorption. When oxygen is available in the plasma, it reacts with the adsorbed silicon-containing species and oxidizes them, leaving a low volatility and less reactive SiO_xCl_y layer at the sidewall. The limiting factor in such a mechanism is generally the oxygen flux. Since oxygen originates from the plasma, this leads to triangular-shaped passivation layers, as explained in Chapter 2. It should be noted that these processes are very sensitive to any oxygen source, including chamber walls and mask sputtering [BEL 96, PAR 05].

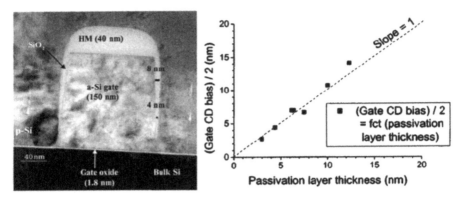

Figure 4.2. *(left) TEM image of a silicon gate etched using a HBr/Cl₂/O₂-based process and (right) correlation between the gate CD bias (defined as the difference in width between the bottom and the top of the gate) and the passivation layer thickness at the top of the gate etched with a HBr/Cl₂/O₂-based plasma (from L. Desvoivres et al. [DES 01] and X. Detter et al. [DET 03])*

For gate patterning with $HBr/Cl_2/O_2$-based plasmas, the etch process is generally split into several steps. The first step involving high ion energy called breakthrough (BT) is used to remove the native oxide from the pSi surface. The main etch (ME) step is then performed to define the upper part of the gate. During this step, the plasma conditions are tuned to obtain the straightest profile, at the expense of the etch selectivity. Then, the soft landing (SL) step is used to finish the etching and land on the gate oxide. A trade-off needs to be found between pattern profile control and selectivity to the gate oxide. Finally, the process is finished with the over etch (OE) step that involves the removal of silicon residues. This step must be infinitely selective to the gate oxide. This sequence enables us to obtain a straight profile and complete etching with a high selectivity in all regions, including dense areas, isolated areas and edges of the active areas with topography due to the isolation. Figure 4.3 shows typical pattern profiles obtained from SEM images after each step for dense and isolated patterns. Clearly, the difficulty lies in maintaining a correct profile simultaneously in all the regions of the wafer.

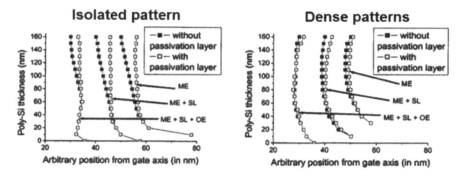

Figure 4.3. *Gate profile obtained from SEM images of pSi gates etched with the main etch (ME) step, main etch and soft landing (ME+SL) steps, and main etch, soft landing and over etch (ME+SL+OE) steps. The gate profile is measured before (solid symbols) and after (open symbols) the removal of passivation layers. The x-axis corresponds to an arbitrary position indicated by a vertical line. (left) For an isolated pattern and (right) for dense patterns (from X. Detter et al. [DET 03])*

4.1.2. *Etching in fluorine-based plasmas*

An alternative process for gate patterning is based on fluorine-containing plasmas. Fluorine-based plasmas spontaneously etch the silicon, leading to isotropic etching. To ensure a vertical gate profile, the plasma must contain etch inhibitors that form passivation layers and block the etching at the pattern sidewalls. Fluorocarbon (FC) gases such as CHF_3 or CH_2F_2 are generally used in combination with SF_6 to obtain anisotropic etching [LUE 11]. These processes are well suited for carbon-based masks that are hardly etched by FC-based plasmas. However, silicon oxide is easily etched with FC-based plasmas, which makes such masks not suitable for fluorine-based plasmas and makes such processes not appropriate for landing on a thin SiO_2-based gate oxide.

During silicon etching with fluorine-based processes with the addition of FC gases, an FC-based reactive layer forms at the etch front that is exposed to ion bombardment [STA 98, VEG 05]. The ion bombardment provides the energy for the reactive fluorine to diffuse towards the interface with silicon and eventually break C–F bonds in the FC film to provide fluorine. It also favors to a lower extent the out-diffusion of etch by-products from the interface to the plasma. The reactions take place at the interface between the FC layer and silicon, consuming fluorine and silicon. The thickness and composition of the reactive layer depend on the plasma gas feedstock and the ion energy, and they control the etching rate when the FC film is thicker than ~1 nm. When the FC layer is thick (low ion bombardment and fluorine-deficient plasma), the etch rate is very slow and limited by the diffusion of radicals and etch by-products. Therefore, the etch rate is very sensitive to the variations of ion bombardment. When the FC is very thin (high ion bombardment and fluorine-rich plasmas), the species easily diffuse and the arrival rate of fluorine limits the etching rate. Therefore, the etch rate is very sensitive to variations of the fluorine density in the plasma [LUE 11].

In addition, the ion bombardment sputters away the carbon-rich FC layer from the etch fronts [STA 98, LUE 11]. This produces low-volatility species that stick on the pattern sidewalls and block the etching. With such a mechanism, the passivation layer precursors originate from the etch front and deposit at the sidewalls of the patterns, close to the etch front [LUE 11, HUB 92]. This results in thin passivation layers with thicknesses that hardly depend on the density of the patterns. Therefore, the micro-uniformity of the process is better than that for $HBr/Cl_2/O_2$-based plasmas.

The mechanisms for silicon etching and sidewall passivation are illustrated in Figure 4.4.

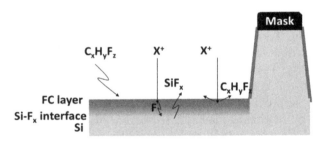

Figure 4.4. *Illustration of silicon etching and sidewall passivation with fluorine-based plasmas with the addition of FC gases. X^+ represents a positive ion of any chemical composition*

4.1.3. *Comparison between the two options*

We have seen that $HBr/Cl_2/O_2$ and SF_6/FC-based plasmas present different etching and passivation mechanisms. The former present a high selectivity towards silicon oxide but have differences between isolated and dense lines because of gas-phase limited passivation. The latter are more suited for carbon-based masks and have a low selectivity with silicon oxide, but provide a better micro-uniformity.

Another advantage of fluorine-based processes is their lower sensitivity towards pSi doping type. Indeed, n^+-doped pSi etches isotopically in chlorine-based plasmas, leading to non-uniformity between N and P gates. Fluorine-based plasmas are less sensitive to the doping level, which also improves uniformity [FLA 90].

The last point to take into account when choosing an etch process is the compatibility with all the gate stack materials. Indeed, for high-k metal gate, the pSi etch process lands on the gate metal. Therefore, the metal will be oxidized by oxygen-containing processes or fluorinated by fluorine-containing processes [LEG 07]. Depending on the metal properties and etch mechanisms, one may be preferred to the other. It may also be interesting to etch the pSi and the metal in one step. In this case, the pSi etch process must etch the metal and not the high-k dielectric [LEG 07]. Again, exposure of the high-k dielectric may lead to fluorination or oxidation that may or may not be acceptable depending on the high-k dielectric material [WU 07, LUE 11].

Table 4.1 presents some comparisons between $HBr/Cl_2/O_2$ and fluorine-based plasma processes. Depending on the process requirement, we may choose one process or the other, or even combine these processes with multiple step processes or by adding FC gases in $HBr/Cl_2/O_2$-based processes.

	$HBr/Cl_2/O_2$-based plasmas	SF_6/FC-based plasmas
Preferred mask material	SiO_2	Carbon-based masks
Selectivity with SiO_2 gate dielectric	Very high	Very low
Sensitivity to gate doping level	Moderate	Low
Sidewall passivation mechanism	Oxidation of silicon-containing products by oxygen radicals coming from the gas phase ➔ high CD micro-loading	Direct deposition of FC species sputtered from the etch front ➔ thin passivation layer and reduced micro-loading
Compatibility with gate metals	Oxygen may oxidize metals, making metal etching difficult if metal oxy-halides are not volatile	Possibility of residues if metal fluorides are not volatile

Table 4.1. *Comparison between $HBr/Cl_2/O_2$ and SF_6/FC-based silicon etching processes for gate patterning*

4.2. Metal gate etching

4.2.1. *Generalities on metal etching requirements*

For high-k metal gate transistor fabrication with the gate-first approach, it is necessary to etch the metal. In some cases, different metals will be used for P and N transistors. In this case, the etch process must be able to pattern N- and P-type gates simultaneously, which increases the requirements in terms of process compatibility and selectivity [YEO 04].

Metal gate patterning has the following four major issues:

– Metal profile: the profile of the metal gate must be as straight as possible. This requirement is mitigated, though, by the low thickness of metal layers integrated in advanced circuits.

– Selectivity: the metal etch process must be selective with the gate oxide to prevent gate oxide damage or removal.

– Chamber contamination: the accumulation of the metal on the chamber walls is a source of process drift. Specific cleaning procedures must be implemented to avoid chamber contamination by metal etch by-products [RAM 09].

– Process compatibility: the metal etch process must be compatible with the whole gate stack. The profile of the pSi gate above the metal must be preserved during metal etching [LEG 07].

There are many different materials potentially integrated as a metal gate in the industry. Among them, titanium nitride is considered to be a standard material that is widely used, eventually in combination with other materials. In the following, we will discuss in more detail the etching of TiN and present very briefly other gate metals.

4.2.2. *TiN etching*

TiN can easily react with oxygen, leading to a thin titanium oxide layer on the top of TiN. This layer may lead to micromasking during TiN etching if it is not removed appropriately [LEG 07]. In addition, oxygen sources must be avoided during TiN etching to avoid the formation of titanium oxide at the etch front, which could also lead to micromasking.

Titanium can be etched using halogen gases including fluorine-, chlorine- and bromine-based gases. The dominant etch products are titanium tetra-halides. Fluorine plasmas are reactive with titanium and can etch titanium nitride spontaneously when the sample temperature is above 50°C [DAR 06]. Since fluorine can also spontaneously etch silicon, it is difficult to obtain a good compatibility between the TiN etch process and the pSi gate profile when fluorine plasmas are used. By adding FC gases, we can improve the protection of pSi gate sidewalls [LUE 11]. Note that FC species do not stick on TiN [DAR 06]. Carbon can also participate in oxygen removal from the TiN surface by forming CO and CO_2. Chlorine-based plasmas are also reactive to titanium and can etch TiN spontaneously. Additional gases must be introduced to avoid micromasking from titanium oxide since Cl-based processes are highly selective to TiO_x [LEG 07]. Bromine-based plasmas are much less reactive with TiN, and titanium etching in HBr-based plasmas is close to physical sputtering and leads to tapered profiles. Combination of HBr and Cl_2 plasma can take advantage of Cl reactivity to obtain straighter sidewalls and of the lower reactivity of Br to prevent lateral etching of TiN and pSi [LEG 07, LUE 11]. Figure 4.5 shows a pSi on TiN gate etched using a multistep $HBr/Cl_2/(O_2)$-based plasma for pSi and TiN etching. Oxygen flow was stopped 10 nm before reaching the TiN [LEG 07].

Figure 4.5. *TEM image of a ~50 nm-wide pSi on TiN gate pattern on HfO_2 etched using a multistep $HBr/Cl_2/(O_2)$-based plasma (from A. LeGouil et al. [LEG 07])*

During titanium etching, Ti-containing species deposit on the chamber walls. Since TiO_x is difficult to etch, it is recommended to avoid any oxygen-containing plasma in the chamber until chamber cleaning is performed. Fluorine- or chlorine-based plasmas (and their combination) have been reported to efficiently clean titanium residues from the chamber walls [LEG 06, CHE 07, RAM 07, RAM 09].

4.2.3. Other metals

Other metals can be used in the metal gate stack. Table 4.2 lists the various potential metals, with typical etching process gases and etch by-products, as well as associated cleaning chemistries and remarks [RAM 09].

Metal	Etching gas	Etch by-products	Cleaning gas	Remarks
Si	$HBr/Cl_2/O_2$ SF_6/FC	$SiCl_2$, $SiCl_4$, $SiBr_2$, $SiBr_4$ SiF_2, SiF_6	SF_6, SF_6/O_2	
Ti, TiN	HBr/Cl_2 SF_6/FC	$TiCl_4$ TiF_4 ($>50°C$)	Cl_2, SF_6	Easily oxidized and oxide is difficult to remove
Ta, TaN	Cl_2/BCl_3	$TaCl_4$	SF_6/Cl_2	TaO_x species are hard to clean
Mo	SF_6 Cl_2/O_2	MoF_6 $MoOCl_4$	Cl_2/O_2	Very fast isotropic etching in fluorine
W	SF_6/O_2 Cl_2/O_2	WF_6, WOF_4, $WOCl_4$	SF_6/O_2	Very fast isotropic etching in fluorine

Table 4.2. *Typical etching gases, etch by-products and cleaning processes for various metal materials potentially integrated in a gate stack*

4.3. Stopping on the gate oxide

4.3.1. Foot/notch formation

During gate etching, a conductor material is etched and straight profiles are targeted. At the end of the etch process, the full conductor stack is etched and the gate dielectric is exposed to the plasma. In order to correct for non-uniformity (from the plasma etching, the gate stack thicknesses and topographic effects) and avoid etch residues, the etch process is always extended a few seconds or tens of seconds after the gate dielectric is exposed. This extra time is called the OE time. As mentioned earlier, the plasma conditions may be changed during this time to obtain milder plasma conditions and a better selectivity, eventually at the expense of pattern profile control. During the OE step/time, less material is etched by the plasma, which reduces the chemical's reactive loading. In addition, an insulator is exposed to the plasma, which leads to the differential charging effect [HWA 97]. For these reasons, the bottom of the gate (where the passivation layers are the thinnest and ions are deflected) may be eroded, eventually leading to a notched profile. Even if a notched profile is not desirable, this effect may be useful to slightly reduce the foot of the gate, as shown in Figure 4.6 [PAR 05].

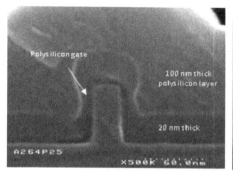 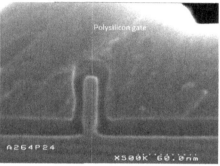

Figure 4.6. *Gate patterns of 75 nm pSi on 1.2 nm of SiO_2 etched with (left) 6s or (right) 8s of the landing step. The pattern etched with the 6s landing step presents a foot while the extra 2s of the landing step transform the foot in the notch. Note that the pattern's lateral dimensions are different and that a specific encapsulation is used to improve the SEM observation (from E. Pargon et al. [PAR 05])*

4.3.2. Selectivity issues with thin SiO_2 and silicon recess

Ion bombardment plays a key role in plasma processing since their energies generate directional chemical ion-induced etching reactions, leading to etch product formation and desorption or even direct sputtering of the material. However, if high energy ions are the key to controlling etch anisotropy, the ions' energy may jeopardize the etch precision required when ultra-thin layers are involved in the patterning of the stack. Today, plasma-induced damage in the source/drain region is a critical point during gate patterning. Indeed, these areas are protected by the gate oxide at the end of the gate etch process, but the gate oxide is so thin that plasma species can diffuse or be implanted through it [DON 99, VAL 99, VIT 03].

It is important to define the difference between etch selectivity and precision of the etch process. Indeed, for example, HBr/O_2 plasma generates etch selectivity between silicon and SiO_2 of at least 100 (i.e. the etch rate of silicon is 100 times faster than that of SiO_2). However, when SiO_2 becomes thin (less than 2 nm), oxidation of silicon below the oxide is observed even if the plasma operating conditions used enable high selectivities on thick SiO_2 layers, leading to important silicon recess after wet cleaning, as shown in Figure 4.7(b) [VIT 03].

It has been demonstrated in several studies [PET 12, VAL 99, DON 99, TUD 01] that high energy ions can get implanted through ultra-thin layers of materials generating damage in the underlying layers (silicon substrate in this case), thereby compromising the precision of the etch process. In other words, when the etching must stop on ultra-thin layers, the concept of etch selectivity loses significance; the most important parameter that drives the etch precision is the ion-induced perturbation thickness in the etched material, which is always several nanometers thick (see Figure 4.7(b)).

As a result, minimizing the so-called silicon recess is extremely difficult to achieve but mandatory. Indeed, this silicon recess/damage directly impacts the performance of the device [ERI 99]. This silicon recess/damage is all the more mandatory and difficult to achieve/control for 3D structures (FinFET or stacked nanowires) since a long over etch (OE > 100%) is required to fully clear the gate, or for planar FDSOI where the underneath SOI must be kept intact (see Figure 1.13, Chapter 1).

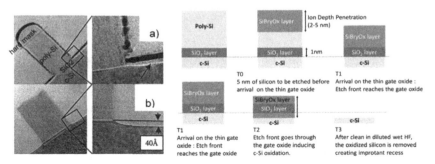

Figure 4.7. *Description of the pSi etch mechanism stopping on SiO_2 layers, leading to c-Si film damage as a function of HBr/O_2 plasma time with associated TEM images of pSi gate after etching a) and followed by a wet cleaning b), leading to important Si recess (from [VIT 03])*

Based on the plasma etch mechanism described previously, ion bombardment control is mandatory to minimize the recess phenomena. However, in conventional plasmas, the lowest energy of ions is around 20 eV (defined by the plasma potential). Under these ion bombardment energy conditions, significant damage on the material exposed to the plasma is still observed [BAR 95]. One approach for improving plasma damage is to lower the electron temperature (low Te) [TIA 06], which reduces the plasma potential and therefore lowers the ion energy to levels below those typically involved in etch processes. Another approach is to use pulsed plasma technologies (allowing a control of plasma dissociation and/or ion energy distribution function) [BAN 12, KAN 12]. However, both technologies already available in the industry are only partially satisfactory since in the first case, low Te plasmas generate broad ion angular distribution functions that reduce the etch directionality, causing dimensional control issues [KAN 12]. The use of pulsed plasma technologies allows a decrease in plasma-induced damage but not their full elimination [PET 10].

The solutions proposed are only partial solutions since a trade-off has to be found between film damage and profile control. Therefore, it is clearly necessary to develop new technologies that will allow the etching of these materials with atomic precision. The ultimate goal would be to selectively etch one layer of atoms at a time (atomic layer etching (ALE)), without disturbing the bulk of the material underneath.

ALE is a technique that removes thin layers of the material using sequential self-limiting reactions. This technique has been developed over 25 years and is today considered one of the most promising techniques for achieving the low process variability described in the previous sections. The ALE cycle, depicted in Figure 4.8, starts with a modification step to form a reactive layer, followed by the modified/reactive layer removal selectively to the non-modified underneath layer.

A variety of self-limiting mechanisms have been widely investigated in the literature and summarized by Kanarik et al. [KAN 15] and Oehrlein et al. [OEH 15]. Thus, different modification mechanisms such as chemisorption [ATH 96], deposition [MET 14] and conversion [POS 13] have been proposed. Removal methods include thermal desorption, neutral beam, ion bombardment, chemical reaction [POS 16] and wet cleaning [POS 13].

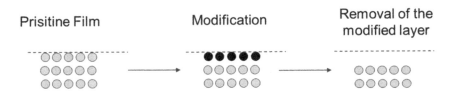

Figure 4.8. *Description of the ALE concept*

This ALE technique is a promising approach for advanced devices requiring precise layer etching and high OE. Indeed, it has been demonstrated that no gate oxide loss and only slight (if any) damage of the bulk silicon underneath the gate oxide is observed after ALE with 30 nm OE (see Figure 4.9) [TAN 15].

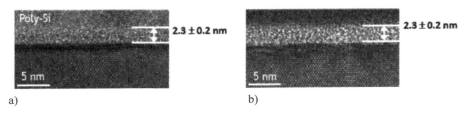

a)   b)

Figure 4.9. *TEM images of gate oxide before a) and after b) ALE etch with an OE amount of about 30 nm [TAN 15]*

4.3.3. *Selectivity/residues with high-k dielectrics*

When the gate etch process stops on a high-k dielectric, the film thickness and resistance to the OE process are larger than for SiO_2, which mitigates the selectivity and silicon recess issues. However, the formation of residues must be taken into account. Indeed, the surface of the high-k dielectric may be modified by the plasma etch process, or residues from the gate etch process may remain on the high-k dielectric [LUE 11]. This is especially the case when a fluorine-containing plasma is used for gate etching. Non-volatile HfF_x species are formed at the surface of the high-k dielectric. These species are very difficult to etch and lead to surface residues after the high-k dielectric etching. In addition, some HfF_x species may be sputtered from the surface and contaminate the chamber walls [RAM 07]. Therefore, it is not recommended to land a fluorine-containing gate etching process on a hafnium-containing high-k dielectric.

4.4. High-k dielectric etching

For high-k metal gate integration, the high-k dielectric must be etched after the gate patterning. Typical high-k dielectrics are hafnium-based dielectrics. The standard process that is adopted in the industry is based on BCl_3-based plasmas [SHA 03]. Such a process tends to deposit a BCl polymer on the surface of the materials exposed to the plasma. Upon ion bombardment, this polymer formation is disfavored and material etching starts. However, the ion bombardment threshold energy between deposition and etching depends on the material exposed to the plasma. Therefore, by wisely tuning the ion energy in a narrow process window, we may obtain polymer deposition on a material (typically the silicon from the source/drain region or the ultra-thin interfacial silicon oxide between the high-k dielectric and the silicon from the source/drain region) while etching another material (typically the high-k dielectric) [SUN 07]. Figure 4.8 shows the etch rate of silicon, silicon oxide and hafnium oxide in a BCl_3 plasma as a function of the plasma bias power (proportional to the ion energy). Negative etch rates correspond to the deposition rate of the BCl polymer. It is clear that hafnium oxide can be etched for lower ion energies than silicon and silicon oxide, which opens a process window (between 4 and 8 W of bias power shown in Figure 4.10) for infinite selectivity. The difference in threshold energy is explained by the formation of Si–B bonds on silicon that favor the polymer deposition while B–O bonds lead to the formation of volatile BOCl on oxygen-containing surfaces [SUN 07]. After the etching process, the BCl polymer can be removed using a wet cleaning process such as diluted HF/HCl [BEC 05] or a dry process such as Cl_2 plasma.

Figure 4.8 also shows the roughness measured by AFM after high-k etching for two bias power conditions. At a low bias power, a large surface roughness is measured, while at a higher bias power, the surface remains smooth. Indeed, when the deposition rate of the BCl polymer on the underlying layer (Si or SiO_2) is larger than the etch rate of the high-k dielectric, some high-k residues may get encapsulated inside the BCl polymer, leading to surface roughness [SUN 09]. Therefore, faster high-k etching than BCl deposition is required, which further narrows down the process window. We can tune the process and eventually enlarge the process window by varying the substrate temperature and adding other gases in the plasma [KIT 06, SUN 07, WAN 08].

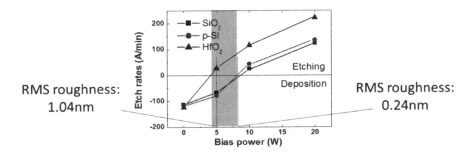

RMS roughness:
1.04nm

RMS roughness:
0.24nm

Figure 4.10. *Etch rate of pSi, silicon oxide and hafnium oxide and deposition rate of the BC polymer on these materials in a BCl₃ plasma as a function of the bias power. The process window with infinite selectivity is between 4 and 8 W here (from E. Sungauer et al. [SUN 09])*

4.5. Line width roughness transfer during gate patterning

For the most advanced technological nodes, the pattern dimensions reach the sub-10 nm range. A very tight profile control is therefore required to avoid too much variability between the devices. However, even if narrow patterns can be formed by lithography and eventually plasma etching, the sidewall roughness (line width roughness (LWR) defined as the 3 sigma standard variation of the line width) is usually above 3 nm, which is too large compared to the pattern dimension. This LWR comes from a stochastic effect during lithography and polymer randomness in the photoresist [MAC 10].

During the gate etching process, the LWR from the photoresist is partially transferred to the gate stack, as shown in Figure 4.11. When the photoresist is exposed to the BARC and SiO_2 hard mask opening steps, we can see that the LWR is strongly reduced. This reduction is explained by the impact of plasma UV photons on the photoresist properties [PAR 09a]. Indeed, energetic UV photons from the plasma lead to photochemical modifications of the photoresist, breaking C=O and C–O–C bonds and eventually releasing ester and lactone groups [PAR 09b]. This changes the photoresist properties and its viscosity. The surface tensions then tend to smoothen the sidewalls of the photoresist and reduce the LWR [PAR 09b]. This photoresist smoothening effect can be advantageously used in specific "resist curing" processes eventually performed by plasma before the gate patterning. The LWR that is transferred into the hard mask is therefore reduced compared to the initial LWR of the photoresist mask. During the gate patterning, the LWR from the hard mask is directly transferred towards the gate. Even if it may be partially masked by the passivation layers, the gate LWR after the full patterning and cleaning steps is identical to the LWR of the mask, as shown in Figure 4.11 [PAR 08]. This transfer of LWR from the hard mask to the gate pattern has been verified for various mask materials and gate stacks [PAR 08].

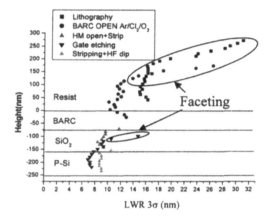

Figure 4.11. *CD-AFM measurement performed after each technological step involved in silicon gate patterning with a SiO₂ hard mask. Each dot corresponds to the LWR measured at a given height of the stack. Mask faceting leads to measurement artifact (from Pargon et al. [PAR 08])*

4.6. Chamber wall consideration after gate patterning

The gate patterning process is generally performed as one process split into many process steps intended to etch the various materials of the gate stack. After the full etch process, the chamber walls are coated with residues that originate from the combination of all the process steps and stack materials. Figure 4.12 shows an analysis of the material residues after each of the individual steps in the case of pSi on TiN gate etching [LEG 06]. In this case, the reactor walls are made of Al_2O_4. After the silicon etching process in $HBr/Cl_2/O_2$-based plasma, the reactor walls are coated with a $SiO_xBr_yCl_z$ layer by the same process as the formation of passivation layers (oxidation of $SiBr_xCl_y$ species at the chamber walls) [CUN 05]. This coating hardly evolves with a slight change in the halogen species concentration during the HBr-based silicon overt etch process. The TiN film is then etched by a short BT step with an argon plasma to remove the native oxide and then by a HBr/Cl$_2$-based plasma. During these steps, some Ti gets deposited on the reactor walls. It is clear that the coating on the chamber walls resulting from the full gate etching process results from a combination of the species deposited during each individual step of the process, eventually modified by the subsequent steps. Note that after hafnium etching, some hafnium may be deposited on the chamber walls, and must be removed using a Cl$_2$-based plasma before using a fluorine-based plasma in the chamber to avoid the formation of hard- to-remove HfF$_x$ species [RAM 08].

To avoid chamber contamination and obtain reproducible process conditions, it is recommended to clean the chamber after each wafer is processed. These cleaning steps can be performed automatically in the chamber between each wafer processing and without the wafer in the chamber. Adequate plasma conditions must be chosen to efficiently remove all residues with the shortest cleaning time [RAM 08, RAM 09]. In addition, the process reproducibility can be further improved by adding a seasoning step before each wafer. This step consists of depositing a thin film of controlled materials (typically SiO$_x$Cl$_y$ by SiCl$_4$/O$_2$ plasma) so that the etching process always starts with perfectly controlled chamber coatings. This layer is removed during the cleaning process.

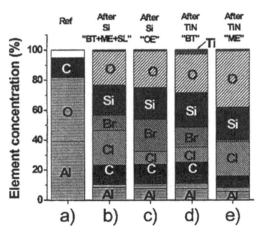

Figure 4.12. *Chemical composition of the coating formed on the reactor wall during pSi/TiN gate patterning a) before process, b) after mask opening and pSi etching, c) after the Si OE, d) after the TiN BT step in an Ar plasma and e) after TiN etching (from A. LeGouil et al. [LEG 06])*

4.7. Summary

Plasma etching plays a key role in the fabrication of high-performance devices. However, with the introduction of new materials and the emergence of complex sub-20 nm device structures involving complex stacks of ultra-thin layers, unprecedented challenges such as critical atomistic scale dimension and profile control are encountered. Now more than ever, plasma etch technology is used to push the limits of semiconductor device fabrication into the nanoelectronics age.

There is a need to develop processes that meet technology requirements: dimension control and film integrity with an atomic precision. This will require improvements in plasma technology (plasma sources, chamber design, etc.), new chemistries (etch gases, flows, interactions with substrates, etc.) as well as a compatibility with new materials used in the gate patterning.

4.8. Bibliography

[ATH 96] ATHAVALE S.D., ECONOMOU D.J., "Realization of atomic layer etching of silicon", *Journal of Vacuum Science and Technology B*, vol. 14, no. 6, p. 3702, 1996.

[BAN 12] BANNA S., ANKARGUL A., CUNGE G. *et al*, "Pulsed high-density plasmas for advanced dry etching processes", *Journal of Vacuum Science and Technology A*, vol. 30, no. 4, p. 040801, 2012.

[BAR 95] BARONE M.E., GRAVES D.B., "Molecular-dynamics simulations of direct reactive ion etching of silicon by fluorine and chlorine", *Journal of Applied Physics*, vol. 78, no. 11, p. 6604, 1995.

[BEC 05] BECKX S., DEMAND M., LOCOROTONDO S. *et al.*, "Implementation of high-k and metal gate materials for the 45 nm node and beyond: gate patterning development", *Microelectronics Reliability*, vol. 45, nos 5–6, pp. 1007–1011, 2005.

[BEL 96] BELL F., JOUBERT O., VALLIER L., "Polysilicon gate etching in high density plasmas. II. X-ray photoelectron spectroscopy investigation of silicon trenches etched using a chlorine-based", *Journal of Vacuum Science and Technology B*, vol. 14, no. 3, pp. 1796–806, 1996.

[BEL 97] BELL F.H., JOUBERT O., "Polysilicon gate etching in high density plasmas. V. Comparison between quantitative chemical analysis of photoresist and oxide masked polysilicon gates etched in HBr/Cl$_2$/O$_2$ plasmas", *Journal of Vacuum Science and Technology B*, vol. 15, no. 1, pp. 88–97, 1997.

[CHE 07] CHEVOLLEAU T., DARNON M. *et al.*, "Analyses of chamber wall coatings during the patterning of ultralow-k materials with a metal hard mask: Consequences on cleaning strategies", *Journal of Vacuum Science and Technology B*, vol. 25, no. 3, p. 886, 2007.

[CUN 02] CUNGE G., INGLEBERT R.L., JOUBERT O. *et al.*, "Ion flux composition in HBr/Cl$_2$/O$_2$ and HBr/Cl$_2$/O$_2$/CF$_4$ chemistries during silicon etching in industrial high-density plasmas", *Journal of Vacuum Science and Technology B*, vol. 20, no. 5, p. 2137, 2000.

[CUN 05] CUNGE G., KOGELSCHATZ M., JOUBERT O. *et al.*, "Plasma–wall interactions during silicon etching processes in high-density HBr/Cl$_2$/O$_2$ plasmas", *Plasma Sources Science and Technology*, vol. 14, no. 2, pp. S42–S52, 2005.

[DAR 06] DARNON M., CHEVOLLEAU T., EON D. *et al.*, "Etching characteristics of TiN used as hard mask in dielectric etch process", *Journal of Vacuum Science and Technology B*, vol. 24, no. 5, p. 2262, 2006.

[DES 00] DESVOIVRES L., VALLIER L., JOUBERT O., "Sub-0.1 μm gate etch processes: towards some limitations of the plasma technology?", *Journal of Vacuum Science and Technology B*, vol. 18, no. 1, p. 156, 2000.

[DES 01] DESVOIVRES L., VALLIER L., JOUBERT O., "X-ray photoelectron spectroscopy investigation of sidewall passivation films formed during gate etch processes", *Journal of Vacuum Science and Technology B*, vol. 19, no. 2, p. 420, 2001.

[DET 03] DETTER X., PALLA R., THOMAS-BOUTHERIN I. *et al.*, "Impact of chemistry on profile control of resist masked silicon gates etched in high density halogen-based plasmas", *Journal of Vacuum Science and Technology B*, vol. 21, no. 5, pp. 2174–2183, 2003.

[DON 99] DONNELLY V.M., KLEMENS F.P., SORSCH T.W. *et al.*, "Oxidation of Si beneath thin SiO$_2$ layers during exposure to HBr/O$_2$ plasmas, investigated by vacuum transfer x-ray photoelectron spectroscopy", *Applied Physics Letters*, vol. 74, no. 9, p. 1260, 1999.

[ERI 99] ERIGUCHI K., NAKAKUBO Y., MATSUDA A. *et al.*, "Plasma-induced defect-site generation in si substrate and its impact on performance degradation in scaled MOSFETs", *IEEE Electron Device Letters*, vol. 30, no. 7, pp. 1275–1277, 2009.

[FLA 90] FLAMM D., "Mechanisms of silicon etching in fluorine-and chlorine-containing plasmas", *Pure and Applied Chemistry*, vol. 62, no. 9, pp. 1709–1720, 1990.

[HUB 92] HÜBNER H., "Calculations on deposition and redeposition in plasma etch processes", *Journal of the Electrochemical Society*, vol. 139, no. 11, p. 3302, 1992.

[HWA 97] HWANG G.S., GIAPIS K.P., "On the origin of the notching effect during etching in uniform high density plasmas", *Journal of Vacuum Science and Technology B*, vol. 15, no. 1, p. 70, 1997.

[KAN 12] KANARIK K.J., KAMARTHY G., GOTTSCHO R.A., "Plasma etch challenges for FinFET transistors", *Solid State Technology*, vol. 55, no. 3, pp. 14–17, 2012.

[KAN 15] KANARIK K., LILL T., HUDSON E. *et al.*, "Overview of atomic layer etching in the semiconductor industry", *Journal of Vacuum Science and Technology A*, vol. 33, no. 2, p. 020802, 2015.

[KIT 06] KITAGAWA T., NAKAMURA K., OSARI K. *et al.*, "Etching of high- k dielectric HfO_2 films in BCl_3 -containing plasmas enhanced with O_2 addition", *Japanese Journal of Applied Physics*, vol. 45, no. 10, pp. L297–L300, 2006.

[LAN 00] LANE J.M., BOGART K.H., KLEMENS F.P. *et al.*, "The role of feedgas chemistry, mask material, and processing parameters in profile evolution during plasma etching of Si(100)", *Journal of Vacuum Science and Technology A*, vol. 18, no. 5, p. 2067, 2000.

[LEG 06] LE GOUIL A., PARGON E., CUNGE G. *et al.*, "Chemical analysis of deposits formed on the reactor walls during silicon and metal gate etching processes", *Journal of Vacuum Science and Technology B*, vol. 24, no. 5, p. 2191, 2006.

[LEG 07] LE GOUIL A., JOUBERT O., CUNGE G. *et al.*, "Poly-Si/TiN/HfO_2 gate stack etching in high-density plasmas", *Journal of Vacuum Science and Technology B*, vol. 25, no. 3, p. 767, 2007.

[LUE 11] LUERE O., PARGON E., VALLIER L. *et al.*, "Etch mechanisms of silicon gate structures patterned in $SF_6/CH_2F_2/Ar$ inductively coupled plasmas", *Journal of Vacuum Science and Technology B*, vol. 29, no. 1, p. 011028, 2011.

[MAC 10] MACK C.A., "Simple model of line edge roughness", *Future Fab Technologies*, vol. 34, 2010.

[MET 14] METZLER D., BRUCE R., ENGELMANN S. *et al.*, "Fluorocarbon assisted atomic layer etching of SiO2 using cyclic Ar/C4F8 plasma", *Journal of Vacuum Science and Technology A*, vol. 32, no. 2, p. 020603, 2014.

[OEH 15] OEHRLEIN G.S., METZLER D., LIA C., "Atomic layer etching at the tipping point: an overview", *ECS Journal of Solid State Science and Technology*, vol. 4, no. 6, pp. N5041–N5053, 2015.

[PAR 05] PARGON E., DARNON M., JOUBERT O. *et al.*, "Towards a controlled patterning of 10 nm silicon gates in high density plasmas", *Journal of Vacuum Science and Technology B*, vol. 23, no. 5, p. 1913, 2005.

[PAR 08] PARGON E., MARTIN M., THIAULT J. *et al.*, "Linewidth roughness transfer measured by critical dimension atomic force microscopy during plasma patterning of polysilicon gate transistors", *Journal of Vacuum Science and Technology B*, vol. 26, no. 3, p. 1011, 2008.

[PAR 09a] PARGON E., MARTIN M., MENGUELTI K. *et al.*, "Plasma impact on 193 nm photoresist linewidth roughness: role of plasma vacuum ultraviolet light", *Applied Physics Letters*, vol. 94, no. 10, p. 103111, 2009.

[PAR 09b] PARGON E., MENGUELTI K., MARTIN M. *et al.*, "Mechanisms involved in HBr and Ar cure plasma treatments applied to 193 nm photoresists", *Journal of Applied Physics*, vol. 105, no. 9, p. 094902, 2009.

[PET 10] PETIT-ETIENNE C., DARNON M., VALLIER L. *et al.*, "Reducing damage to Si substrates during gate etching processes by synchronous plasma pulsing", *Journal of Vacuum Science and Technology B*, vol. 28, no. 5, p. 926, 2010.

[PET 12] PETIT-ETIENNE C., PARGON E., DAVID S. *et al.*, "Silicon recess minimization during gate patterning using synchronous plasma pulsing", *Journal of Vacuum Science and Technology B*, vol. 30, no. 4, p. 040604, 2012.

[POS 13] POSSEME N., POLLET O., BARNOLA S., "Alternative process for thin layer etching: Application to nitride spacer etching stopping on silicon germanium", *Applied Physics Letters*, vol. 105, p. 051605, 2014.

[POS 16] POSSEME N., AH-LEUNG V., POLLET O. *et al.*, "Thin layer etching of silicon nitride: A comprehensive study of selective removal using NH3/NF3 remote plasma", *J. Vac. Sci. Technol. A*, vol. 34 no. 6, p. 061301, 2016.

[RAM 07] RAMOS R., CUNGE G., JOUBERT O. *et al.*, "Plasma/reactor walls interactions in advanced gate etching processes", *Thin Solid Films*, vol. 515, no. 12, pp. 4846–4852, 2007.

[RAM 08] RAMOS R., CUNGE G., JOUBERT O., "Plasma reactor dry cleaning strategy after TiN, TaN and HfO$_2$ etching processes", *Journal of Vacuum Science and Technology B*, vol. 26, no. 1, p. 181, 2008.

[RAM 09] RAMOS R., CUNGE G., JOUBERT O. *et al.*, "Plasma reactor dry cleaning strategy after TaC, MoN, WSi, W, and WN etching processes", *Journal of Vacuum Science and Technology B*, vol. 27, no. 1, p. 113, 2007.

[SHA 03] SHA L., PUTHENKOVILAKAM R., LIN Y.-S. *et al.*, "Ion-enhanced chemical etching of HfO_2 for integration in metal–oxide–semiconductor field effect transistors", *Journal of Vacuum Science and Technology B*, vol. 21 no. 6, p. 2420, 2004.

[STA 98] STANDAERT T.E.F.M., SCHAEPKENS M., RUEGER N.R. *et al.*, "High density fluorocarbon etching of silicon in an inductively coupled plasma: Mechanism of etching through a thick steady state fluorocarbon layer", *Journal of Vacuum Science and Technology A*, vol. 16, no. 1, p. 239, 1998.

[SUN 07] SUNGAUER E., PARGON E. *et al.*, "Etching mechanisms of HfO_2, SiO_2, and poly-Si substrates in BCl_3 plasmas", *Journal of Vacuum Science and Technology B*, vol. 25, p. 1640, 2007.

[SUN 09] SUNGAUER E., MELLHAOUI X., PARGON E. *et al.*, "Plasma etching of HfO_2 in metal gate CMOS devices", *Microelectronics Engineering*, vol. 86, nos. 4–6, pp. 965–967, 2009.

[TAN 15] TAN S., YANG W., KANARIK K.J. *et al.*, "Highly selective directional atomic layer etching of silicon", *ECS Journal of Solid State Science and Technology*, vol. 4, no. 6, pp. N5010–N5012, 2015.

[TIA 06] TIAN C., NOZAWA T., ISHIBASI K. *et al.*, "Characteristics of large-diameter plasma using a radial-line slot antenna", *Journal of Vacuum Science and Technology A*, vol. 24, no. 4, p. 1421, 2006.

[TUD 01] TUDA M., SHINTANI K., TANIMURA J., "Study of plasma–surface interactions during overetch of polycrystalline silicon gate etching with high-density HBr/O2 plasmas", *Applied Physics Letter*, vol. 79, p. 2535, 2001.

[VAL 99] VALLIER L., DESVOIVRES L., BONVALOT M. *et al.*, "Thin gate oxide behavior during plasma patterning of silicon gates", *Applied Physics Letters*, vol. 75, no. 8, p. 1069, 1999.

[VEG 05] VÉGH J.J., HUMBIRD D., GRAVES D.B., "Silicon etch by fluorocarbon and argon plasmas in the presence of fluorocarbon films", *Journal of Vacuum Science and Technology A*, vol. 23, no. 6, p. 1598, 2005.

[VIT 03] VITALE S.A., SMITH B.A., "Reduction of silicon recess caused by plasma oxidation during high-density plasma polysilicon gate etching", *Journal of Vacuum Science and Technology B*, vol. 21, no. 5, p. 2205, 2004.

[WAN 08] WANG C., DONNELLY V.M., "Mechanisms and selectivity for etching of HfO$_2$ and Si in BCl$_3$ plasmas", *Journal of Vacuum Science and Technology* A, vol. 26, no. 4, p. 597, 2008.

[WU 07] WU W.C., LAI C.S., WANG J.C. *et al.*, "High-performance HfO$_2$ gate dielectrics fluorinated by postdeposition CF$_4$ plasma treatment", *Journal of the Electrochemical Society*, vol. 154, no. 7, p. H561, 2007.

[YEO 04] YEO Y.-C., "Metal gate technology for nanoscale transistors—material selection and process integration issues", *Thin Solid Films*, vols 462–463, pp. 34–41, 2004.

List of Acronyms

ALE: Atomic Layer Etching

ARC: Anti-Reflective Coating

ARDE: Aspect Ratio-Dependent Etching

BARC: Bottom Anti-Reflective Coating

BCP: Block Copolymer Materials

BOX: Buried OXide

BT: BreakThrough

CAIBE: Chemically Assisted Ion Beam Etching

CCP: Capacitively Coupled Plasmas

CD: Critical Dimension

CDU: Critical Dimension Uniformity

CMOS: Complementary Metal Oxide Semiconductor

COO: Cost Of Ownership

CVD: Chemical Vapor Deposition

CW: Continuous Wave

DC: Direct Current

DIBL: Drain-Induced Barrier Lowering

DOE: Design of Orthogonal Experiments

DOF:	Depth Of Focus
DP:	Double Patterning
DSA:	Direct Self-Assembly
EBDW:	E-Beam Direct Write
E-beam:	Electron-beam
ECR:	Electron Cyclotron Resonance
EOT:	Equivalent Oxide Thickness
EUVL:	Extreme UltraViolet Lithography
FC:	FluoroCarbon
FDSOI:	Fully Depleted Silicon On Insulator
FI:	Factory Interface
FIB:	Focus Ion Beam
FOUP:	Front Opening Unified Pods
HP:	Half Pitch
HVM:	High Volume Manufacturing
IBE:	Ion Beam Etching
ICP:	Inductively Coupled Plasmas
IEDF:	Ion Energy Distribution Function
ITRS:	International Technology Roadmap for Semiconductors
LED:	Light Emitting Diodes
LELE:	Lithography-Etch, Lithography-Etch
LER:	Line Edge Roughness
LWR:	Line Width Roughness
ME:	Main Etch
MEMS:	Micro-Electro-Mechanical Systems
MOSFET	Metal–Oxide–Semiconductor Field-Effect Transistor
NA:	Numerical Aperture

NIL:	NanoImprint Lithography
NTRS:	National Technology Roadmap for Semiconductors
OE:	Over Etch
OPL:	Organic Planarizing Layer
PECVD:	Plasma Enhanced Chemical Vapor Deposition
PMMA:	PolyMethyl Meth Acrylate
PR:	PhotoResist
PS:	PolyStyrene
RF:	Radio Frequency
RIBE:	Reactive Ion Beam Etching
RLSA:	Radial Line Slot Antenna
RMG:	Replacement Metal Gate
RSD:	Raised Source–Drain
SADP:	Self-Aligned Double Patterning
SAQP:	Self-Aligned Quadruple Patterning
SCE:	Short Channel Effect
SEM:	Scanning Electron Microscope
SIA:	Semiconductor Industry Association
Si-ARC:	Si-Anti Reflective Coating
SIMS:	Secondary Ion Mass Spectrometry
SL:	Soft Landing
SOC:	Spin On Carbon
TCP:	Transformer Coupled Plasma
UV:	UltraViolet
VUV:	Vacuum UV

List of Authors

Sébastien BARNOLA
CEA-LETI-Minatec
Grenoble
France

Maxime DARNON, Ph.D.
Laboratoire des Nanotechnologies
et Nanosystèmes
CNRS
Sherbrooke
Canada

Stefan LANDIS, Ph.D.
CEA-LETI-Minatec
Grenoble
France

Nicolas POSSEME, Ph.D.
CEA-LETI-Minatec
Grenoble
France

Maud VINET, Ph.D.
CEA-LETI-Minatec
Grenoble
France

Index

Printed in the United States
By Bookmasters